Introduction to Atmospheric Ch
fundamentals of atmospheric ch
it reviews our basic understanding of the chemistry of the Earth's atmosphere and some outstanding environmental issues, including air pollution, acid rain, the ozone hole, and global change.

Peter Hobbs is an eminent atmospheric science teacher, researcher, and author of several well-known textbooks. This text and Hobbs' other Cambridge University Press book, *Basic Physical Chemistry for the Atmospheric Sciences* (second edition, 2000), form ideal companion volumes for a full course in atmospheric chemistry. Subjects covered include evolution of the Earth's atmosphere; interactions between solar and terrestrial radiation and atmospheric chemical species; sources, transformations, transport, and sinks of chemicals in the atmosphere; atmospheric gases and particles; cloud and precipitation chemistry; biogeochemical cycling; air pollution; and stratospheric chemistry. Student exercises are provided at the end of each chapter.

The book is designed to be a primary textbook for a first university course (undergraduate or graduate) in atmospheric chemistry and will be adopted in departments of atmospheric science, meteorology, environmental science, geophysics, and chemistry. It is also eminently suitable for self-instruction.

Professor Peter V. Hobbs (University of Washington) is known internationally for his research on many aspects of the atmosphere: clouds, precipitation, aerosols, storms, atmospheric chemistry, and climate. He is the author of the definitive text *Ice Physics* (Oxford University Press), the author of *Basic Physical Chemistry for the Atmospheric Sciences* (Cambridge University Press), coauthor (with J. M. Wallace) of one of the most widely used textbooks in meteorology, *Atmospheric Sciences: An Introductory Survey* (Academic Press), and editor of several other books. He has authored more than 300 scientific papers. Professor Hobbs has served on many national and international committees, including the Scientific Steering Committee of the International Global Atmospheric Chemistry Program. He has been a visiting senior research scientist in England, France, Germany, and Italy.

INTRODUCTION TO ATMOSPHERIC CHEMISTRY

A Companion Text to *Basic Physical Chemistry for the Atmospheric Sciences*

PETER V. HOBBS
University of Washington

PUBLISHED BY THE PRESS SYNDICATE OF THE UNIVERSITY OF CAMBRIDGE
The Pitt Building, Trumpington Street, Cambridge, United Kingdom

CAMBRIDGE UNIVERSITY PRESS
The Edinburgh Building, Cambridge CB2 2RU, UK http://www.cup.cam.ac.uk
40 West 20th Street, New York, NY 10011-4211, USA http://www.cup.org
10 Stamford Road, Oakleigh, Melbourne 3166, Australia
Ruiz de Alarcón 13, 28014 Madrid, Spain

First published 2000

Printed in the United States of America

Typeface Times Roman 10/13 pt. *System* QuarkXPress™ [BTS]

A catalog record for this book is available from the British Library.

Library of Congress Cataloging in Publication Data

Hobbs, Peter Victor
Introduction to atmospheric chemistry / Peter V. Hobbs.
p. cm.
Includes bibliographical references.
ISBN 0-521-77143-9 (hb)
1. Atmospheric chemistry. I. Title: Atmospheric chemistry. II. Title.

QC879.6 .H62 2000
551.51′1 – dc21

99-053320

ISBN 0 521 77143 9 hardback
ISBN 0 521 77800 X paperback

Contents

Preface *page* ix

1 Evolution of the Earth's atmosphere 1
1.1 The primitive atmosphere 2
1.2 Prebiotic atmosphere and the origins of life 3
1.3 Rise of oxygen and ozone 5
1.4 Oxygen and carbon budgets 6
1.5 Some other atmospheric constituents 9
1.6 The Gaia hypothesis 10
1.7 Summary 10

2 Half-life, residence time, and renewal time of chemicals in the atmosphere 13
2.1 Half-life 13
2.2 Residence time and renewal time 15
2.3 Spatial and temporal scales of variability 20

3 Present chemical composition of the atmosphere 21
3.1 Units for chemical abundance 21
3.2 Composition of air close to the Earth's surface 23
3.3 Change in atmospheric composition with height 26

4 Interactions of solar and terrestrial radiation with atmospheric trace gases and aerosols 33
4.1 Some basic concepts and definitions 34
4.2 Attenuation of solar radiation by gases 41
4.3 Vertical profile of absorption of solar radiation in the atmosphere 43

4.4 Heating of the atmosphere due to gaseous absorption of solar radiation 45
4.5 Attenuation of solar radiation by aerosols 50
4.6 Absorption and emission of longwave radiation 51
4.7 The greenhouse effect, radiative forcing, and global warming 54
4.8 Photochemical reactions 57

5 Sources, transformations, transport, and sinks of chemicals in the troposphere 63
5.1 Sources 63
5.2 Transformations by homogeneous gas-phase reactions 72
5.3 Transformations by other processes 78
5.4 Transport and distributions of chemicals 79
5.5 Sinks of chemicals 80

6 Atmospheric aerosols 82
6.1 Aerosol concentrations and size distributions 82
6.2 Sources of aerosols 91
6.3 Transformations of aerosols 95
6.4 Chemical composition of aerosols 97
6.5 Transport of aerosols 99
6.6 Sinks of aerosols 100
6.7 Residence times of aerosols 102
6.8 Geographical distribution of aerosols 104
6.9 Atmospheric effects of aerosols 104

7 Cloud and precipitation chemistry 111
7.1 Overview 111
7.2 Cloud condensation nuclei and nucleation scavenging 113
7.3 Dissolution of gases in cloud droplets 121
7.4 Aqueous-phase chemical reactions 125
7.5 Precipitation scavenging 131
7.6 Sources of sulfate in precipitation 134
7.7 Chemical composition of rainwater 135
7.8 Production of aerosols by clouds 137

8 Tropospheric chemical cycles 143
8.1 Carbon cycle 143
8.2 Nitrogen cycle 149
8.3 Sulfur cycle 151

9 Air pollution 153
9.1 Sources of anthropogenic pollutants 153
9.2 Some atmospheric effects of air pollution 156

10 Stratospheric chemistry 164
10.1 Unperturbed stratospheric ozone 165
10.2 Anthropogenic perturbations to stratospheric ozone 171
10.3 Stratospheric aerosols; sulfur in the stratosphere 179

Appendix I Exercises 185
Appendix II Answers to exercises in Appendix I and hints and solutions to the more difficult exercises 206
Appendix III Atomic weights 235
Appendix IV The International System of Units (SI) 238
Appendix V Some useful numerical values 240
Appendix VI Suggestions for further reading 241

Index 242

Preface

This short book is a companion volume and a natural extension to my textbook entitled *Basic Physical Chemistry for the Atmospheric Sciences* (Cambridge University Press, 1995; second edition published in 2000). Together these two books provide material for a first (undergraduate or graduate) course in atmospheric chemistry; they should also be suitable for self-study.

In *Basic Physical Chemistry for the Atmospheric Sciences* the groundwork was laid for courses in atmospheric chemistry and other areas of environmental chemistry. The present book provides a short introduction to the subject of atmospheric chemistry itself. Twenty years ago this subject was a minor branch of the atmospheric sciences, pursued by relatively few scientists. Today, atmospheric chemistry is one of the most active and important disciplines within meteorology, and one with which every geoscientist and environmental scientist should have some familiarity.

The emphasis of this book is on the basic principles of atmospheric chemistry, with applications to such important environmental problems as air pollution, acid rain, the ozone hole, and global change. In keeping with the pedagogical approach of its companion volume, model solutions are provided to a number of exercises within the text. In an appendix, readers are invited to test their skills on further exercises. Answers to all of the exercises and worked solutions to the more difficult ones, are provided.

Thanks are due to Halstead Harrison for allowing me to use some of his exercises, and to Richard Gammon, Dean Hegg, Daniel Jaffe, Robert Kotchenruther, Conway Leovy, Donald Stedman, and Stephen Warren for reviewing various portions of this book. I thank also the National Science Foundation and the National Aeronautics and Space Administration for their support of my own research on atmospheric chemistry.

Comments on this book, which will be gratefully received, can be sent by e-mail to phobbs@atmos.washington.edu. Current information on the book, including any errata, can be found on http://cargsun2.atmos.washington.edu/~phobbs/IntroAtmosChem/Info.html.

Peter V. Hobbs
Seattle

1

Evolution of the Earth's atmosphere

The composition of the Earth's atmosphere is unique within the solar system. The Earth is situated between Venus and Mars, both of which have atmospheres consisting primarily of CO_2 (an oxidized compound);[1,a] the outer planets (Jupiter, Saturn, Uranus, Neptune) are dominated by reduced compounds, such as CH_4. By contrast, CO_2 and CH_4 are only minor (although very important) constituents of the Earth's atmosphere. Nitrogen represents ~78% of the molecules in air, and life-sustaining oxygen accounts for ~21%. The presence of so much oxygen is surprising, since it might appear to produce a combustible mixture with many of the other gases in air (e.g., sulfur to form sulfates, nitrogen to form nitrates, hydrogen to form water).

The Earth's atmosphere is certainly not in chemical equilibrium, since the concentrations of N_2, O_2, CH_4, N_2O, and NH_3 are much higher than they would be for perfect equilibrium. Why is this so? A clue is provided by Table 1.1, which lists the five most common elements in the Earth's atmosphere, biosphere, hydrosphere, crust, mantle, and core. Four of the most abundant elements in the atmosphere (nitrogen, oxygen, hydrogen, and carbon) are also among the top five most abundant elements in the biosphere. This suggests that biological processes have played a dominant role in the evolution of the Earth's atmosphere and that they are probably responsible for its present chemical nonequilibrium state. However, as we will see, this has occurred in relatively recent times. In this chapter we will speculate on the development of the Earth's atmosphere since it was first formed some 4.5 billion years ago (4.5 Ga), at which time it probably had no (or very little) atmosphere.

[a] Numerical superscripts in the text (1, 2, . . . etc.) refer to **Notes** at the end of the chapter.

Table 1.1. *The five most abundant elements (in terms of the number of atoms) in the major chemical reservoirs on Earth (the numbers in parentheses are the masses, in kg, of the reservoirs)*[a]

Atmosphere (5.2×10^{18})	Biosphere[b] (4.2×10^{15})	Hydrosphere[c] (2.4×10^{21})	Crust (2.4×10^{22})	Mantle (4.0×10^{24})	Core[d] (1.9×10^{24})
N	H	H	O	O	Fe
O	O	O	Si	Si	Ni
H	C	Cl	Al	Mg	C
Ar	N	Na	Fe	Fe	S
C	Ca	Mg	Mg/Ca	Al	Si

[a] Adapted from P. Brimblecombe, *Air Composition and Chemistry*, Cambridge University Press, Cambridge, 1996, p. 4.
[b] Includes plants, animals, and organic matter but not coal or sedimentary carbon.
[c] Water in solid and liquid form on or above the Earth's surface.
[d] Composition of Earth's core is uncertain.

1.1 The primitive atmosphere

In comparison to the Sun (or the cosmos) the atmosphere of the Earth is deficient in the light volatile elements (e.g., H) and the noble or inert gases (e.g., He, Ne, Ar, Kr, Xe). This suggests that either these elements escaped as the Earth was forming or the Earth formed in such a way as to systematically exclude these gases (e.g., by the agglomeration of solid materials similar to that in meteorites[2]). In either case, the Earth's atmosphere was probably generated by the degassing of volatile compounds contained within the original solid materials that formed the Earth (a so-called *secondary atmosphere*).

Earlier models of the evolution of the Earth hypothesized that it formed relatively slowly with an initially cold interior that was subsequently heated by radioactive decay. This would have allowed gases to be released by volcanic activity. Until the Earth's core formed, these gases would have been highly reducing (e.g., H_2, CH_4, NH_3), but after the formation of the core they would have been similar to the effluents from current volcanic activity (i.e., H_2O, CO_2, N_2, and small quantities of H_2, CO, and sulfur compounds). More recent models suggest that the Earth's interior was initially hot due to tremendous bombardment (a major impact during this period formed the Moon). In this case, the Earth's core would have formed earlier and volcanic gases emitted 4.5 Ga ago could have been similar to present emissions (i.e., more oxidized). Also, many of the volatile materials could have been released by the impacts themselves, resulting in an atmosphere of steam during the period that the Earth was accreting material.

When the accretionary phase ended and the Earth cooled, the steam could have condensed and rained out to produce the oceans. The atmosphere that was left would likely have been dominated by CO_2, CO, and N_2.[3] The partial pressure of CO_2 and CO in the primitive atmosphere could have been ~10 bar,[4] together with ~1 bar from nitrogen. The Earth continued to be bombarded, even after the main accretionary period, until at least 3.8 Ga ago. If these impacts were cometary in nature, they could have provided CO (by oxidation of organic carbon or by reduction of atmospheric CO_2 by iron-rich impactors) and NO (by shock heating of atmospheric CO_2 and N_2).

1.2 Prebiotic atmosphere and the origins of life

Life on Earth is unlikely to have started (or at least to have survived) during the period of heavy bombardment. However, the fossil record shows that primitive forms of living cells were present no later than 3.5 Ga ago. Laboratory experiments demonstrate that many biologically important organic compounds, including amino acids that are basic to life, can form when a mixture of CH_4, NH_3, H_2, and H_2O is irradiated with ultraviolet (UV) light or sparked by an electric discharge (simulating lightning). However, CH_4 and NH_3 may not have been present 3.5 Ga ago unless the oxidation state of the upper mantle, which affects the chemical composition of volcanic effluents, differed from its present composition. Even if CH_4 and NH_3 were released from volcanoes, they would have been only minor atmospheric constituents because they are quickly photolysed. Thus, the early atmosphere was probably dominated by N_2 and CO_2 (with a concentration perhaps 600 times greater than at present), with trace amounts of H_2, CO, H_2O, O_2, and reduced sulfur gases (i.e., a "weakly reducing" atmosphere). Due to the photodissociation of CO_2

$$CO_2 + h\nu \rightarrow CO + O$$

where $h\nu$ represents a photon of frequency ν, followed by

$$O + O + M \rightarrow O_2 + M$$

where M represents an inert molecule that can remove some of the energy of the reaction, molecular oxygen would have increased sharply with altitude above ~20 km because of the increased intensity of solar radiation. The concentrations of O_2 at the surface would have been very low ($<10^{-12}$ present atmospheric levels, PAL) due, in part, to reactions with H_2.

Two key compounds for the formation of life are probably formaldehyde (HCHO) and hydrogen cyanide (HCN), which are needed for the synthesis of sugars and amino acids, respectively. Formaldehyde could have formed by photochemical reactions involving N_2, H_2O, CO_2, H_2, and CO (removal of HCHO from the atmosphere by precipitation would have provided a source of organic carbon for the oceans). Formation of HCN, from N_2 and CO_2, for example, is much more difficult because it requires breaking the strong triple bonds of N_2 and CO. This can occur in lightning discharges, but the N and C atoms are more likely to combine with atomic oxygen than with each other unless $[C]/[O] > 1$. It is because of this difficulty that theories have been invoked involving the introduction of biological precursor molecules by comets and the origins of life in oceanic hydrothermal vents.

Exercise 1.1. A catalytic cycle that might have contributed to the formation of H_2 from H in the early atmosphere of the Earth is

$$H + CO + M \xrightarrow{k_1} HCO + M \quad \text{(i)}$$

$$H + HCO \xrightarrow{k_2} H_2 + CO \quad \text{(ii)}$$

$$\text{Net:} \quad 2H \rightarrow H_2$$

If this cycle were in steady state, and if the concentrations of CO and M were 1.0×10^{12} and 2.5×10^{19} molecule cm^{-3}, respectively, and the magnitudes of the rate coefficients k_1 and k_2 are 1.0×10^{-34} cm^6 s^{-1} molecule^{-2} and 3.0×10^{-10} cm^3 s^{-1} molecule^{-1}, respectively, what would have been the concentration of the radical HCO?

Solution. The rate of formation of HCO by Reaction (i) is $k_1[H][CO][M]$, where the square brackets indicate concentrations in molecules per cm^3. The rate of destruction of HCO by Reaction (ii) is $k_2[H][HCO]$. At steady state, the rate of formation of HCO must equal its rate of destruction. Therefore,

$$k_1[H][CO][M] = k_2[H][HCO]$$

or

$$[HCO] = \frac{k_1}{k_2}[CO][M]$$
$$= \frac{1.0 \times 10^{-34}}{3.0 \times 10^{-10}}(1.0 \times 10^{12})(2.5 \times 10^{19})$$
$$\simeq 8.3 \times 10^6 \, molecule \, cm^{-3}$$

Well-founded astrophysical theory leads us to believe that the temperature of the Sun has increased since its birth to the present time. Thus, 4.6 Ga ago the Sun was probably 25% to 30% weaker than it is now (the so-called faint young Sun). If the early atmosphere had a chemical composition similar to the present, its equilibrium surface temperature with respect to the faint young Sun would have been below 0°C until about 2 Ga ago. However, the formation of sedimentary rocks ~3.8 Ga ago, and the development of life which started more than 3.5 Ga ago, indicate that liquid water was present at these early times. Since CO_2 is a "greenhouse" gas (i.e., it reduces the loss of longwave radiation to space from the Earth's surface), its presence in high concentrations in the Earth's early atmosphere could have maintained the temperature of the Earth above freezing some ~3.5 to 3.8 Ga ago even with a faint young Sun.

Cooling of the Earth might have triggered a negative feedback involving CO_2 and the chemical weathering of rocks. For example, in addition to the $CaCO_3$ reservoir, dissolved CO_2 reacts with rhodochrosite ($MnCO_3(s)$),[5]

$$MnCO_3(s) + CO_2(g) + H_2O(l) \rightleftarrows Mn^{2+}(aq) + 2HCO_3^-(aq)$$

But with decreasing temperature, this and other similar sinks for CO_2 decrease, thereby allowing atmospheric CO_2 concentrations to increase.

1.3 Rise of oxygen and ozone

The advent of biological activity on Earth led the way to rapid increases in atmospheric molecular oxygen through photosynthesis. In photosynthesis by green plants, light energy is used to convert H_2O and CO_2 into O_2 and energy-rich organic compounds called carbohydrates (e.g., glucose, $C_6H_{12}O_6$), which are stored in the plants

$$6H_2O(l) + 6CO_2(g) + h\nu \rightarrow 6O_2(g) + C_6H_{12}O_6(s) \qquad (1.1)$$

Exercise 1.2. What change in the oxidation number of the carbon atom is produced by Reaction (1.1)?

Solution. Since the oxidation number of each oxygen atom in CO_2 is –2, the oxidation number of the C atom is +4. In $C_6H_{12}O_6$ the oxidation numbers of the H and O are +1 and –2, respectively. Therefore, the oxidation number of the C atom in $C_6H_{12}O_6$ is 0. Hence, Reaction (1.1) decreases the oxidation number of the C atom from +4 to zero; that is, the carbon is reduced. (Note that the reverse of Reaction (1.1) will oxidize the C atom, since its oxidation number will rise from zero to +4.)

The geologic record shows that atmospheric O_2 first reached appreciable concentrations ~2 Ga ago. The combined atmosphere-ocean system appears to have gone through three main stages. In the first stage, almost the entire system was a reducing environment. In the next stage the atmosphere and the surface of the ocean presented an oxidizing environment, although the deep ocean was still reducing. In the third (and current) stage, the entire system is oxidizing with abundant free molecular oxygen (O_2).

The earliest life forms probably developed in aqueous environments, far enough below the surface to be protected from the Sun's lethal UV radiation but close enough to the surface to have access to visible solar radiation needed for photosynthesis. There is also speculation that life might have originated in hydrothermal systems in the deep ocean, where bacteria do not rely on photosynthesis.

By means of processes to be discussed in Section 10.1, the buildup of oxygen in the atmosphere led to the formation of the ozone layer in the upper atmosphere, which filters out UV radiation from the Sun. With the development of the ozone layer, less and less UV radiation reached the Earth's surface. In this increasingly favorable environment, plant life was able to spread to the uppermost layers of the ocean, thereby gaining access to increasing amounts of visible radiation, an essential ingredient in the photosynthesis Reaction (1.1). More oxygen – less UV radiation – more access to visible radiation – more abundant plant life – still more oxygen production: through this bootstrap process, life may have slowly but inexorably worked its way upward toward the surface until it finally emerged onto land some 400 million years ago.

1.4 Oxygen and carbon budgets

For every molecule of oxygen produced in Reaction (1.1), one atom of carbon is incorporated into an organic compound. Most of these carbon

Table 1.2. *Estimate of inventory of carbon near the Earth's surface (units are gigatons (10^{15} g) of carbon)*

Biosphere:	
Marine	2–5
Terrestrial (land, plants)	600
Atmosphere (as CO_2)	750
Ocean (as dissolved CO_2)	38,000
Fossil fuels	8,000
Shales	8,000,000
Carbonate rocks	65,000,000

atoms are oxidized in respiration or in the decay of organic matter, which is the reverse of Reaction (1.1). However, for every few tens of thousands of molecular carbons formed by photosynthesis, one escapes oxidation by being buried or "fossilized." Most of the Earth's unoxidized carbon is contained in shales, and smaller amounts are stored in more concentrated forms in fossil fuels (coal, oil, and natural gas). The relatively "short-term" storage of organic carbon in the biosphere represents a minute fraction of the total storage. More quantitative information on the relative amounts of carbon stored in various forms is given in Table 1.2.

The burning of fossil fuels undoes the work of photosynthesis by oxidizing that which was reduced. At the present rate of fuel consumption, humans burn in one year what it took photosynthesis ~1,000 years to produce! This rate of consumption seems less alarming when one bears in mind that photosynthesis has been at work for hundreds of millions of years. One can take further comfort from the fact that the bulk of the organic carbon in the Earth's crust is stored in a form that is far too dilute for humans to exploit.

Of the net amount of oxygen that has been produced by plant life during the Earth's history (i.e., production by photosynthesis minus consumption by respiration and the decay of organic matter), only about 10% is presently stored in the atmosphere. Most of the oxygen has found its way into oxides (such as Fe_2O_3) and biogenically precipitated carbonate compounds ($CaCO_3$ and $CaMg(CO_3)_2$) in the Earth's crust. The biological formation of carbonate compounds is of particular interest since it is the major sink for the vast amounts of CO_2 that have been released in volcanic activity.

Carbonates are formed by means of ion exchange reactions that take place within certain marine organisms, the most important being the one-celled foraminifera. The dissolved CO_2 forms a weak solution of carbonic acid (H_2CO_3)

$$H_2O + CO_2 \rightleftarrows H_2CO_3 \rightleftarrows H^+ + HCO_3^- \quad (1.2)$$

It has been suggested that a sequence of reactions then follows, the net result of which is

$$2(HCO_3^-) + Ca^{2+} \rightleftarrows CaCO_3 + H_2CO_3 \quad (1.3)$$

The $CaCO_3$ enters into the shells of animals, which fall to the sea floor and are eventually compressed into limestone in the Earth's crust. The hydrogen ions released in Reaction (1.2) react with metallic oxides in the Earth's crust, from which they steal an oxygen atom to form another water molecule. The stolen oxygen atom is eventually replaced by one from the atmosphere. Thus, oxygen is removed from the atmosphere during the formation of carbonates, and it is given back to the atmosphere when carbonates dissolve. It has been proposed that foraminifera and other carbonate-producing sea species, by virtue of their role as mediators in the process of carbonate formation, regulate the amount of oxygen present in the atmosphere, which has been remarkably constant over the past few million years.

The widespread occurrence of marine limestone deposits suggests that ion exchange reactions in sea water have played an important role in the removal of CO_2 from the Earth's atmosphere. Therefore, the dominance of CO_2 in the present Martian atmosphere may be due, in part at least, to the absence of liquid water on the surface. In contrast to the situation on Mars, the massive CO_2 atmosphere of Venus may be a consequence of the high surface temperatures on that planet. At such temperatures there should exist an approximate state of equilibrium between the amount of CO_2 in the atmosphere and the carbonate deposits in rocks on the surface, as expressed by the reaction

$$CaCO_3(s) \rightleftarrows CaO(s) + CO_2(g) \quad (1.4)$$

The concentration of CO_2 in the Earth's atmosphere has been rising steadily since the early part of this century (Fig. 1.1), which suggests that the rate of removal of CO_2 from the Earth's atmosphere is not large enough to keep pace with the ever-increasing rate of input due to the burning of fossil fuels. However, the present rate of increase in atmo-

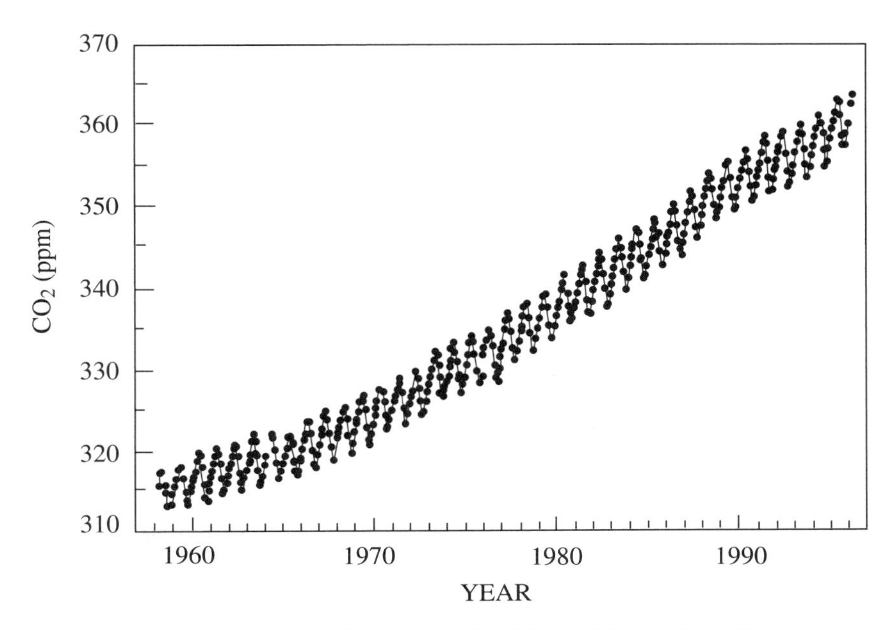

Figure 1.1. Concentration of atmospheric CO_2 at Mauna Loa Observatory, Hawaii, for the period 1958–1996. Data prior to May 1974 are from the Scripps Institute of Oceanography, and data since May 1974 are from the National Oceanic and Atmospheric Administration.

spheric CO_2 is only about half the rate at which CO_2 is being added to the atmosphere by the burning of fossil fuels. This implies that about half of the CO_2 added by fossil fuel burning is going into the oceans, forests, or other sinks.

1.5 Some other atmospheric constituents

By means of ion exchange reactions analogous to Reaction (1.3) and fixation by soil microorganisms, a small fraction of the nitrogen released into the atmosphere has entered into nitrates in the Earth's crust. However, because of the chemical inertness of nitrogen and its low solubility in water (1/70th that of CO_2), most of the nitrogen released by volcanoes has remained in the atmosphere. Because of the nearly complete removal from the Earth's atmosphere of water vapor (to form liquid water in the oceans and hydrated crystalline rocks) and CO_2 by the processes described earlier, nitrogen has become the dominant gaseous constituent of the Earth's atmosphere.

Sulfur and its compounds H_2S and SO_2, which are released into the Earth's atmosphere by volcanic emissions, are quickly oxidized to SO_3, which dissolves in cloud droplets to form a dilute solution of H_2SO_4. After being scavenged from the atmosphere by precipitation particles, the sulfate ions combine with metal ions to form sulfates within the Earth's crust. Sulfur dioxide may also react with NH_3 in the presence of liquid water and an oxidant to produce ammonium sulfate $(NH_4)_2SO_4$.

1.6 The Gaia hypothesis

As we have seen in Section 1.3, the biosphere is responsible for the buildup and maintenance of oxygen in the Earth's atmosphere and for the present nonequilibrium state of the atmosphere. In the *Gaia* ("mother Earth") hypothesis, the influence of the biosphere on the atmosphere is seen as "purposeful." The biosphere and atmosphere are viewed as an ecosystem, in which the chemical composition and climate of the Earth are maintained in optimum states (for the biosphere) by the metabolism and evolutionary development of the biota. This might be achieved through a rich web of positive and negative feedbacks. For example, we saw in Section 1.4 that carbonate-producing sea species might regulate the amount of oxygen in the atmosphere.

Like many stimulating viewpoints, the Gaia hypothesis is controversial. The Darwinian theory, whereby biota adapt to the environment imposed on them, is the more commonly held view, although, as discussed earlier, the atmosphere has been completely reformulated by biological activity.

1.7 Summary

The Earth's primitive atmosphere was probably formed by the accretion of extraterrestrial volatile materials and by outgassing of the Earth's interior. As accretion diminished and the Earth evolved, the steamy atmosphere condensed to form oceans, leaving an atmosphere dominated by CO_2 (~1 to 10 bar), CO and N_2 (~1 bar). Despite a faint young Sun, the initially high concentration of CO_2 maintained surface temperatures on Earth above 0°C by means of the greenhouse effect (see Section 4.7). The weakly reducing primitive atmosphere was favorable for the emergence of biota. Photosynthesis then increased oxygen concentrations, which, in turn, allowed ozone formation in the upper atmosphere by photochemical reactions. The shielding of the Earth's surface from dangerous solar UV radiation by ozone in the upper atmosphere permitted life to evolve onto land. At the same time, the concentrations of CO_2 (and other greenhouse gases) declined, thereby compensating for an increasingly bright Sun. The relatively stable climate of the Earth over the past 3.5 Ga, during which time the mean surface temperature has remained in the range of ~5 to 50°C, is probably due to the negative feedback between surface temperature, atmospheric CO_2, and the weathering rates of rocks.

The likely general trends of O_2, O_3, and CO_2 since the Earth's atmosphere first formed are shown in Figure 1.2.

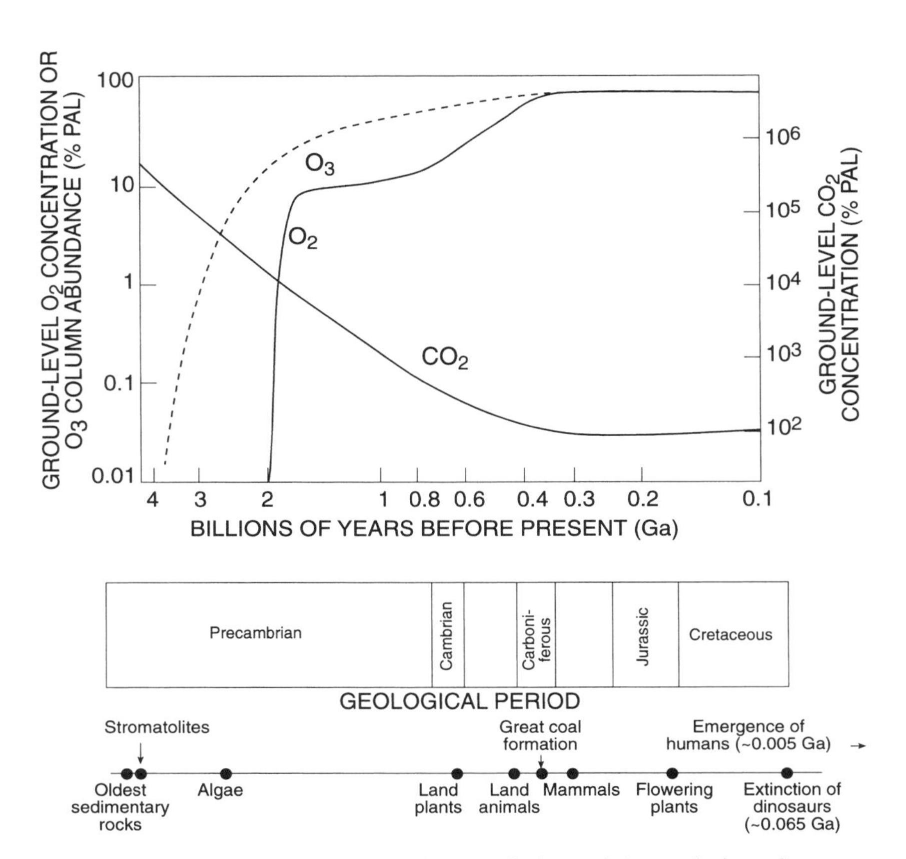

Figure 1.2. Schematic diagram showing predictions of the evolution of oxygen, ozone and carbon dioxide to present atmospheric levels (PAL). [After R. P. Wayne, *Chemistry of Atmospheres*, Oxford University Press, p. 404 (1991) by permission of Oxford University Press; and, J. F. Kasting, personal communication (1999).]

Exercises

See Exercises 1(a)–(f), and Exercises 2–5 in Appendix I.

Notes

1 A list of chemical symbols is given in Appendix III.
2 Such material probably included small amounts of volatile substances (i.e., materials capable of existing in gaseous form within the range of temperatures found on the surface of the Earth). For example, water could have been present as ice or in chemical combination with other solid substances.
3 Carbon-containing compounds are second only to water as the most abundant volatiles on the Earth's surface. However, most carbon on Earth is "tied up" in carbonate rocks. The amount of carbon in the Earth's crust is $\sim 10^{20}$ kg; if all of this were present in the atmosphere as CO_2, the pressure at the Earth's surface would be 60 to 80 times greater than present atmospheric levels (as it is in the atmosphere of Venus).
4 1 bar = 10^5 Pa. 1 mb = 10^2 Pa = 1 hPa. The pressure at the Earth's surface at the present time (1 atmosphere) is ~1.013 bar, or 1,013 hPa.
5 When we wish to emphasize the phase of a chemical species, we will use parenthetical insertions: *g* for gas, *l* for liquid, *s* for solid, and *aq* for aqueous.

2

Half-life, residence time, and renewal time of chemicals in the atmosphere

In atmospheric chemistry it is important to have some idea and some measure of the characteristic times that various chemicals spend in the atmosphere. In this chapter we discuss several ways of doing this. We also discuss a connection between the residence time of a chemical in the atmosphere and its spatial variability.

2.1 Half-life

Let us start by considering a chemical A, which is depleted at a rate that is proportional to its concentration [A] at time t; that is,

$$-\frac{d[\mathrm{A}]}{dt} = k[\mathrm{A}] \tag{2.1}$$

where k is a constant. Then,

$$\int_{[\mathrm{A}]_0}^{[\mathrm{A}]_t} \frac{d[\mathrm{A}]}{[\mathrm{A}]} = -k\int_0^t dt$$

where $[\mathrm{A}]_0$ and $[\mathrm{A}]_t$ are the initial concentrations of A and the concentration of A at time t, respectively. Hence,

$$\ln\frac{[\mathrm{A}]_t}{[\mathrm{A}]_0} = -kt$$

or, converting to base-10 logarithms (indicated by "log"),

$$\log[\mathrm{A}]_t = -\frac{kt}{2.303} + \log[\mathrm{A}]_0 \tag{2.2}$$

The *half-life* ($t_{1/2}$) of a chemical in the atmosphere is defined as the time required for its concentration decrease to half of its initial value.

We can derive an expression for $t_{1/2}$ for the case considered earlier by substituting $[A]_t = [A]_0/2$ and $t = t_{1/2}$ into Eq. (2.2), which yields

$$t_{1/2} = \frac{2.303 \log 2}{k}$$

Therefore,

$$t_{1/2} = \frac{0.693}{k} \tag{2.3}$$

Note that, in this case, $t_{1/2}$ is independent of the initial concentration of A.

A first-order chemical reaction in one reactant A is described by Eq. (2.1), where k is called the *rate coefficient* for the reaction. Because the decay rate per unit mass of a radioactive material (e.g., as measured by the number of clicks per minute of a Geiger counter) is proportional to the number of radioactive atoms present in the remaining sample, its decay is also represented by Eq. (2.1). Radiocarbon dating of organic materials is based on this principle. Carbon-12 (i.e., carbon with a mass number (the number of protons plus the number of neutrons) of 12) is the stable isotope of carbon. Carbon-14 is unstable (i.e., radioactive) with a half-life of 5,700 a. Because carbon-14 is produced by cosmic ray bombardments in the upper atmosphere, the ratio of carbon-14 to carbon-12 in the atmosphere is nearly constant (and is believed to have been so for at least 50,000 a). Carbon-14 is incorporated into atmospheric CO_2, which is in turn incorporated, through photosynthesis, into plants. When animals eat plants, the carbon-14 is incorporated into their tissues. While a plant or animal is alive it has a constant intake of carbon compounds, and it maintains a ratio of carbon-14 to carbon-12 that is identical to that of the air. When a plant or animal dies, it no longer ingests carbon compounds, and the ratio of carbon-14 to carbon-12 decreases with time, due to the radioactive decay of carbon-14. Hence, the period that elapsed since a plant or animal or organic material was alive can be deduced by comparing the ratio of carbon-14 to carbon-12 in the material with the corresponding ratio for air.

Exercise 2.1. A wooden carving, found on an archaeological site, is subjected to radiocarbon dating. The carbon-14 activity is 12.0 counts per minute per gram of carbon, compared to 15.0 counts per minute per gram of carbon for a living tree. What is the maximum age of the carving?

Solution. Since the half-life ($t_{1/2}$) of carbon-14 is 5,700 a, we can substitute this value into Eq. (2.3) to obtain a value for k

$$k = \frac{0.693}{t_{1/2}} = \frac{0.693}{5,700} = 1.22 \times 10^{-4} a^{-1} = 2.32 \times 10^{-10} \text{ min}^{-1}$$

The amount of radioactive carbon-14 (as measured, say, by a Geiger counter) per gram of carbon in the carving at the time the tree from which it was made died (say, at $t = 0$) is given by Eq. (2.1) as

$$[\text{A}]_0 = \frac{(\text{decay rate at } t = 0)}{k}$$

$$= \frac{15.0}{2.33 \times 10^{-10}} = 6.44 \times 10^{10} \textit{ counts per gram of carbon}$$

Similarly, at the present time (say, t years after the tree died), the amount of carbon-14 per gram of carbon in the carving is given by

$$[\text{A}]_t = \frac{12.0}{2.33 \times 10^{-10}} = 5.15 \times 10^{10} \textit{ counts per gram of carbon}$$

Substituting these values of $[\text{A}]_0$, $[\text{A}]_t$, and k (in units of a^{-1}) into Eq. (2.2) yields

$$\log(5.15 \times 10^{10}) = \frac{-(1.22 \times 10^{-4})t}{2.303} + \log(6.44 \times 10^{10})$$

or

$$10.71 = -5.3 \times 10^{-5} t + 10.81$$

Therefore,

$$t = 1.89 \times 10^{3} a$$

Because the carving could not have been made before the tree died, the maximum age of the carving is 1,890 years.

2.2 Residence time and renewal time

Chemicals are injected continually into the atmosphere from natural and anthropogenic sources, and they are also produced by chemical reactions in the air. Yet the overall chemical composition of the atmosphere does not change greatly over relatively short periods of time (although, as we shall see, there are important exceptions). This is because there are sinks that remove trace chemicals from the atmosphere at about the same rate as the chemicals are injected into (and/or produced within) the atmosphere, so that most chemicals in air exist in roughly steady-state conditions.

An important parameter related to a chemical under steady-state conditions is its *residence time*, or *lifetime*, (τ) in the atmosphere, which is defined as

$$\tau = \frac{M}{F} \tag{2.4}$$

where M is the amount of the chemical in the atmosphere and F the *efflux* (i.e., rate of removal plus rate of destruction) of the chemical from the atmosphere. If M and F change with time

$$\tau_t = \frac{M_t}{F_t} \tag{2.5}$$

where the subscript t indicates the value at time t. We can define, in an analogous way, the residence time in terms of the *influx* (i.e., rate of input plus rate of production) of a chemical to the atmosphere.

A useful analogy here is a tank of water, which can represent the atmosphere. Suppose the tank is full of water and is overflowing at its top due to water being pumped into the bottom of the tank at a rate F. Then the rate of removal of water from the tank is F. If we assume that the water entering the bottom of the tank steadily displaces the water above it by pushing it upwards without any mixing, the time spent by each small element of water that enters the bottom of the tank before it overflows at the top is M/F, where M is the volume of the tank (this is the reason for defining residence time as M/F in Eq. (2.4)). In this case, when no mixing occurs in the reservoir, the residence time of the water is the same as the *renewal time* (T), which is defined as the *time required to completely displace the original water from the tank.* That is,

$$\tau = T \quad \text{(for no mixing)} \tag{2.6}$$

Consider now a more realistic situation in which mixing takes place between the material that is injected into the atmosphere and the material already residing in it. For simplicity, we still consider the mixing to be complete and thorough (i.e., *perfect mixing*). The tank analogy is again helpful. Suppose that at time zero the tank is full of dirty water, and at this time clean water starts to be pumped into the bottom of the tank. Since the mixing is perfect, the rate of removal of dirty water from the top of the tank will be proportional to the fraction of the water in the tank that is dirty. Therefore, if W is the amount of dirty water in the tank at time t,

$$-\frac{dW}{dt} = kW \tag{2.7}$$

where k is a constant of proportionality. Since Eqs. (2.7) and (2.1) have the same form, the half-life of the dirty water is given by Eq. (2.3). Also, from Eqs. (2.4) and (2.7) we have for the dirty water

$$\tau = \frac{M}{F} = \frac{W}{(-dW/dt)} = \frac{1}{k} \tag{2.8}$$

Combining Eqs. (2.3) and (2.8), we obtain the following relationship between the half-life ($t_{1/2}$) and the residence time (τ)

$$t_{1/2} = 0.693\,\tau \quad \text{(for perfect mixing)} \tag{2.9}$$

In the case of perfect mixing, the renewal time (T) is strictly infinitely long because some molecules of dirty water will always be present in the tank. However, we can obtain an idea of the "effective" value of T for perfect mixing as follows. From the definition of the half-life ($t_{1/2}$), we know that after $t_{1/2}$ minutes one half of the dirty water will be left in the tank, and after $2t_{1/2}$ minutes $(1/2)(1/2) = (1/2)^2$ of the dirty water will be left in the tank, and so on. Therefore, after $6t_{1/2}$ minutes, $(1/2)^6 = (1/64)$ of the dirty water will be left in the tank. If we (arbitrarily) decide that 1/64 is a sufficiently small fraction that most of the dirty water can be considered to have been displaced, then, for a chemical that is perfectly mixed in the atmosphere and for which the efflux is given by a first-order Reaction (Eq. 2.7), we have the following relationships between the effective renewal time (T), the half-life ($t_{1/2}$) and the residence time (τ)

$$T \simeq 6t_{1/2} \simeq 4\,\tau \quad \text{(for perfect mixing)} \tag{2.10}$$

In practice, of course, the atmosphere falls somewhere between the cases of no mixing and perfect mixing.

In the atmosphere, the very stable gas nitrogen has a residence time of ~1 to 10 million years, whereas oxygen has a residence time of ~3,000 to 10,000 a. The very reactive species sulfur dioxide and water, on the other hand, have residence times in the atmosphere of only a few days and ten days, respectively. Of course, residence times may be determined by physical removal processes (e.g., scavenging by precipitation) as well as chemical processes. The residence times of some gases in the atmosphere are given in Table 2.1.

Table 2.1. *Residence times of some atmospheric gases*[a] *(in many cases only very rough estimates are possible)*

Gas	Residence Time
Nitrogen (N_2)	1.6×10^7 a
Helium (He)	10^6 a
Oxygen (O_2)	3,000–10,000 a
Carbon dioxide (CO_2)	3–4 a
Nitrous oxide (N_2O)	150 a
Methane (CH_4)	9 a
CFC-12 (CF_2Cl_2)	>80 a
CFC-11 ($CFCl_3$)	~80 a
Hydrogen (H_2)	4–8 a
Methyl chloride (CH_3Cl)	2–3 a
Carbonyl sulfide (COS)	~2 a
Ozone (O_3)	100 days
Carbon disulfide (CS_2)	40 days
Carbon monoxide (CO)	~60 days
Water vapor[b]	~10 days
Formaldehyde (CH_2O)	5–10 days
Sulfur dioxide (SO_2)	1 day
Ammonia + Ammonium ($NH_3 + NH_4^+$)	2–10 days
Nitrogen dioxide (NO_2)	0.5–2 days
Nitrogen oxide (NO)	0.5–2 days
Hydrogen chloride (HCl)	4 days
Hydrogen sulfide (H_2S)	1–5 days
Hydrogen peroxide (H_2O_2)	1 day
Dimethyl sulfide (CH_3SCH_3)	0.7 days

[a] The residence time (or lifetime) is defined as the amount of the chemical in the atmosphere divided by the rate at which the chemical is removed from the atmosphere. This time scale characterizes the rate of adjustment of the atmospheric concentration of the chemical if the emission rate is changed suddenly.

[b] The residence time of liquid water in clouds is ~6 h.

Exercise 2.2. Ammonia (NH_3), nitrous oxide (N_2O) and methane (CH_4) comprise 1×10^{-8}, 3×10^{-5}, and 7×10^{-5}% by mass of the Earth's atmosphere, respectively. If the effluxes of these chemicals from the atmosphere are 5×10^{10}, 1×10^{10}, and 4×10^{11} kg a^{-1}, respectively, what are the residence times of NH_3, N_2O, and CH_4 in the atmosphere? (Mass of the Earth's atmosphere = 5×10^{18} kg.)

Solution. From Eq. (2.4) the residence time is given by

$$\tau = \frac{M}{F}$$

where M is the quantity of chemical in the atmosphere, and F the efflux. For NH_3,

$$M = \frac{1 \times 10^{-8}}{100}(5 \times 10^{18})\ \text{kg}$$

and $F = 5 \times 10^{10}\,\text{kg a}^{-1}$, therefore, $\tau_{NH_3} = 0.01\,\text{a} = 4$ days. For N_2O,

$$M = \frac{(3 \times 10^{-5})}{100} \times (5 \times 10^{18})\ \text{kg}$$

and $F = 1 \times 10^{10}\,\text{kg a}^{-1}$, therefore, $\tau_{N_2O} = 150\,\text{a}$. For CH_4,

$$M = \frac{(7 \times 10^{-5})(5 \times 10^{18})}{100}\ \text{kg}$$

and $F = 4 \times 10^{11}\,\text{kg a}^{-1}$, therefore, $\tau_{CH_4} = 9\,\text{a}$.

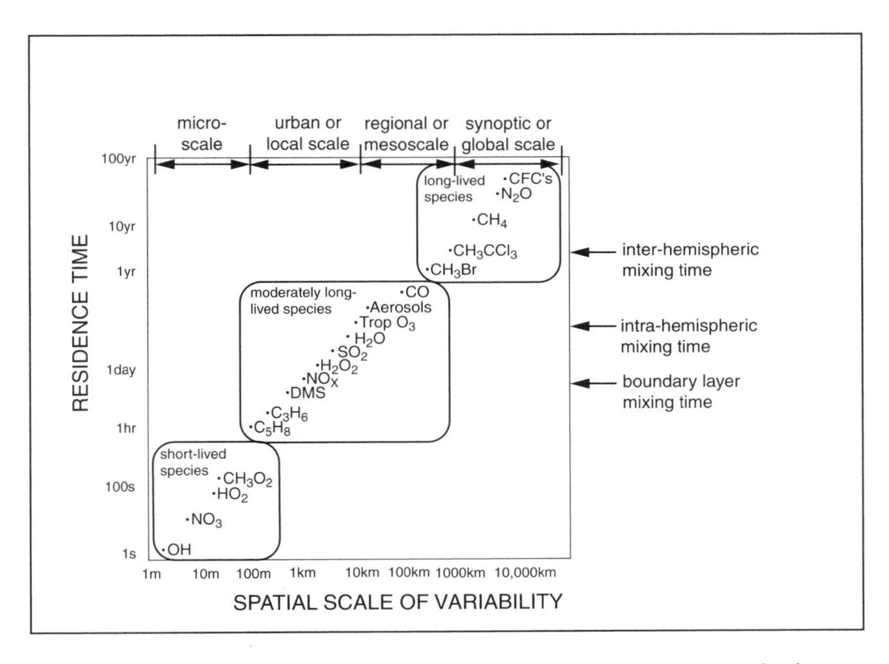

Figure 2.1. Spatial and temporal scales of variability for some atmospheric constituents. The temporal scale is represented by residence time. [Adapted with permission from *The Atmospheric Sciences Entering the Twenty-First Century*. Copyright © 1998 by the National Academy of Sciences. Courtesy of the National Academy Press.]

2.3 Spatial and temporal scales of variability

If a chemical species has a very short (or very long) residence time in the atmosphere, significant variations in the concentration of the species will generally occur over very short (or very large) spatial scales (Fig. 2.1). Species with short residence times will be present in high concentrations close to localized sources, and in low concentrations far removed from their sources.

Exercises

See Exercise 1(g) and Exercises 6–9 in Appendix I.

3
Present chemical composition of the atmosphere

3.1 Units for chemical abundance

Before discussing the chemical composition of the atmosphere, we need to describe the various units that can be used for expressing the amounts (or abundances) of chemicals in the atmosphere.

The most common unit for expressing the abundance of gases in air is *mixing ratio by volume*, that is, the fraction of the volume of the air occupied by a gas. Depending on the magnitude of the mixing ratio, the most convenient units can be *percent by volume*, *parts per million by volume* (ppmv), *parts per billion by volume* (ppbv), or *parts per trillion by volume* (pptv), where, 1 ppmv = 1 unit of volume per 10^6 units (or 10^{-6}), 1 ppbv = 1 unit of volume per 10^9 units (or 10^{-9}), and 1 pptv = 1 unit of volume per 10^{12} units (or 10^{-12}). Note that we use the American definitions of billion (10^9) and trillion (10^{12}). It is very convenient to express the abundance of a gas as a mixing ratio by volume, because, from the gas equation, we know that the volumes occupied by different gases at the same temperature and pressure are proportional to the numbers of molecules of the gases. For example, if the mixing ratio of N_2O in air is 330 ppbv, we know that the fraction of the total number of molecules in air that are N_2O (that is the *mole fraction* of NO_2) is 330×10^{-9}, or, for every 10^9 molecules in the air 330 are N_2O. Also, the partial pressure exerted by a gas in a mixture is proportional to the mole fraction of the gas. Therefore, at the surface of the Earth, where the total pressure of air is 1 atm, N_2O with a mixing ratio of 330 ppbv would exert a partial pressure of 330×10^{-9} atm.

Problem 3.1. If the mixing ratio of CO_2 in the atmosphere is 354 ppmv, how many CO_2 molecules are there in 1 m^3 of air at 1 atm and 0°C?

Solution. Let us calculate first the number of molecules in $1\,m^3$ of any gas at 1 atm and 0°C (which is called *Loschmidt's number*). This can be derived from the ideal gas equation in the form

$$p = n_0 kT \qquad (3.1)$$

where p is the pressure of the gas, n_0 the number of molecules per unit volume, k Boltzmann's[1] constant ($= 1.381 \times 10^{-23}\,J\,deg^{-1}$ molecule^{-1}), and T temperature (in K). When $p = 1\,atm = 1{,}013 \times 10^2\,Pa$, and $T = 273\,K$ are substituted into Eq. (3.1), n_0 is Loschmidt's number. Therefore,

$$\text{Loschmidt's number} = \frac{1{,}013 \times 10^2}{(1.381 \times 10^{-23})\,273}$$
$$= 2.7 \times 10^{25} \text{ molecules m}^{-3}$$

Since the volumes occupied by gases at the same temperature and pressure are proportional to the numbers of molecules in the gases, we can write

$$\frac{\text{volume occupied by CO}_2\text{ molecules in air}}{\text{volume occupied by air}}$$
$$= \frac{\text{number of CO}_2\text{ molecules in 1 m}^3\text{ of air}}{\text{total number of molecules in 1 m}^3\text{ of air}}$$

The left side of this last relation is equal to 354 ppmv. Therefore,

$$354 \times 10^{-6} = \frac{\text{number of CO}_2\text{ molecules in 1 m}^3\text{ of air}}{2.7 \times 10^{25}}$$

Hence, the number of CO_2 molecules in $1\,m^3$ of air is $(354 \times 10^{-6})(2.7 \times 10^{25}) = 9.6 \times 10^{21}$.

The abundances of condensed materials in air, or the amount of cloud water, is generally expressed in terms of mass of material per unit volume of air ($kg\,m^{-3}$ in SI units, but $\mu g\,m^{-3}$ are often used for aerosols and $g\,m^{-3}$ for cloud water). These units are sometimes used for gases, but in this case they are inconvenient because even if two gases are present in the same amount when expressed in, say, $\mu g\,m^{-3}$, the fraction of the total volume of air they occupy may differ because of their different molecular weights. Also, if care is not taken, confusion can arise when the amount of a gas, say SO_2, is given in units of mass of sulfur per m^3 of air ($\mu g(S)\,m^{-3}$) rather than the mass of SO_2.

The abundance of a chemical species may also be expressed by its *mole fraction* (or *mole ratio*), which is the number of moles of the species to the total number of moles of all species in the sample. The units are mol per mol; for very dilute mixtures, more practical units can be used (e.g., nmol/mol). Since moles are proportional to numbers of molecules, the mole fraction is the same as the molecular fraction (in terms of numbers). Therefore, a gas having a mole fraction of $1\,nmol\,mol^{-1}$ would have a mixing ratio of 1 ppbv.

The amount of a solute in a liquid solution is usually expressed as the number of moles of the solute in 1 liter of the solution (i.e., mole L^{-1}, or M for short). This is called the *molar concentration* (or *molarity*) of the solution. Low concentrations in aqueous solution are often given in units such as ppm (parts per million). However, in this case ppm is mass per mass (such as $mg\,kg^{-1}$, which is equivalent to $mg\,L^{-1}$).

3.2 Composition of air close to the Earth's surface

As far as the first few major gaseous constituents are concerned, air is essentially a homogeneous mixture up to an altitude of about 100 km; this region is called the *homosphere*. Table 3.1 lists the major gaseous components of air in the homosphere and their typical concentrations within a few kilometers of the Earth's surface. The two main constituents of air in this region are molecular nitrogen (~78% by volume) and molecular oxygen (~21% by volume). These are followed by water vapor (up to ~4% by volume) and then, a long way behind, by argon (0.93% by volume) and CO_2 (0.036% by volume or 360 ppmv, but growing by ~1.5 ppmv each year).

Most of the constituents of air that are of prime importance in atmospheric chemistry are present in much smaller concentrations, consequently, they are called *trace constituents*. Some of the trace constituents in air are listed in Table 3.1. Of particular interest are reactive species, such as O_3, SO_2, and CO. There are many other reactive trace constituents of major importance (e.g., the radicals[2] OH, HO_2, and Cl) that are present in such low concentrations in air (sub-pptv) that they are difficult to measure even with the most sensitive instruments available. This book is concerned primarily with the reactive trace constituents in the homosphere. However, in the remainder of this chapter a brief description is

Table 3.1. *Composition of clean (nonurban) tropospheric air*[a]

Chemical Species	Concentration[b]	Source
A. *Major and Minor Gases*		
Nitrogen (N_2)	78.08% (780,840 ppmv)	Volcanic, biogenic
Oxygen (O_2)	20.95% (209,460 ppmv)	Biogenic
Argon (Ar)	0.93% (9,340 ppmv)	Radiogenic
Water vapor (H_2O)	Variable – up to 4% (40,000 ppmv)	Volcanic, evaporation
Carbon dioxide (CO_2)	0.036% (355 ppmv)	Volcanic, biogenic, anthropogenic
B. *Trace Constituents*		
1. *Oxygen Species*		
Ozone (O_3)	0–100 ppbv	Photochemical
Atomic oxygen (O) (ground state)	0–10^3 cm^{-3}	Photochemical
Atomic oxygen (O) (O* – excited state)	0–10^{-2} cm^{-3}	Photochemical
2. *Hydrogen Species*		
Hydrogen (H_2)	560 ppbv	Photochemical, biogenic
Hydrogen peroxide (H_2O_2)	10^9 cm^{-3}	Photochemical
Hydroperoxyl radical (HO_2)	0–10^8 cm^{-3}	Photochemical
Hydroxyl radical (OH)	0–10^6 cm^{-3}	Photochemical
Atomic hydrogen (H)	0–1 cm^{-3}	Photochemical
3. *Nitrogen Species*		
Nitrous oxide (N_2O)	310 ppbv	Biogenic, anthropogenic
Nitric acid (HNO_3)	0–100 pptv	Photochemical
Ammonia (NH_3)	0–0.5 ppbv	Biogenic, anthropogenic
Hydrogen cyanide (HCN)	~200 pptv	Anthropogenic (?)
Nitrogen dioxide (NO_2)	0–300 pptv	Photochemical
Nitric oxide (NO)	0–300 pptv	Anthropogenic, biogenic, lightning, photochemical
Nitrogen trioxide (NO_3)	0–100 pptv	Photochemical
Peroxyacetyl nitrate (PAN) ($CH_3CO_3NO_2$)	0–50 pptv	Photochemical
Dinitrogen pentoxide (N_2O_3)	0–1 pptv	Photochemical
Pernitric acid (HO_2NO_2)	0–0.5 pptv	Photochemical
Nitrous acid (HNO_2)	0–0.1 pptv	Photochemical
Nitrogen aerosols:		
Ammonium Nitrate (NH_4NO_3)	~10 pptv	Photochemical
Ammonium chloride (NH_4Cl)	~0.1 pptv	Photochemical
Ammonium sulfate ($(NH_4)_2SO_4$)	~0.1 pptv	Photochemical

Table 3.1. *(cont.)*

Chemical Species	Concentration[b]	Source
4. *Carbon Species*		
Methane (CH_4)	1.7 ppmv	Biogenic, anthropogenic
Carbon monoxide (CO)	70–200 ppbv (N. hemisphere) 40–60 ppbv (S. hemisphere)	Anthropogenic, biogenic, photochemical
Formaldehyde (CH_2O)	0.1 ppbv	Photochemical
Methyl hydroperoxide radical (CH_3OOH)	0–10^{11} cm^{-3}	Photochemical
Methylperoxy radical (CH_3O_2)	0–10^{8} cm^{-3}	Photochemical
Methyl radical (CH_3)	0–10^{-1} cm^{-3}	Photochemical
5. *Sulfur Species*		
Carbonyl sulfide (COS)	0.05 ppbv	Volcanic, anthropogenic, biogenic
Dimethyl sulfide (DMS) ($(CH_3)_2S$)	70–200 pptv	Biogenic
Hydrogen sulfide (H_2S)	0–0.5 ppbv	Biogenic, anthropogenic
Sulfur dioxide (SO_2)	0.2 ppbv	Volcanic, anthropogenic, photochemical
Dimethyl disulfide ($(CH_3)_2S_2$)	5–10 pptv	Biogenic
Carbon disulfide (CS_2)	5–10 pptv	Volcanic, anthropogenic, biogenic
Sulfuric acid (H_2SO_4)	0–20 pptv	Photochemical
Sulfurous acid (H_2SO_3)	0–20 pptv	Photochemical
Sulfur monoxide (SO)	0–10^{3} cm^{-3}	Photochemical
Thiohydroxyl radical (HS)	0–1 cm^{-3}	Photochemical
Sulfur trioxide (SO_3)	0–10^{-2} cm^{-3}	Photochemical
6. *Halogen species*		
Hydrogen chloride (HCl)	1 ppbv	Sea salt, volcanic
Methyl chloride (CH_3Cl)	0.5 ppbv	Biogenic, anthropogenic
Methyl bromide (CH_3Br)	10 pptv	Biogenic, anthropogenic
Methyl iodide (CH_3I)	1 pptv	Biogenic
7. *Noble Gases (chemically inert)*		
Neon (Ne)	18 ppmv	Volcanic (?)
Helium (He)	5.2 ppmv	Radiogenic
Krypton (Kr)	1 ppmv	Radiogenic
Xenon (Xe)	90 ppbv	Radiogenic

[a] Adapted from J. S. Levine in *Global Ecology*, Eds. M. B. Rambler et al., Academic Press, New York, 1989, p. 53.

[b] Typical values at 1 atm are given; many of the trace gases have highly variable concentrations. In addition to percentage by volume, the units are:

parts per million by volume (ppmv) = 10^{-6}
parts per billion by volume (ppbv) = 10^{-9}
parts per trillion by volume (pptv) = 10^{-12}
number density of molecules at the surface (cm^{-3}).

given of the changes in the composition of the atmosphere with increasing altitude above 100 km.

3.3 Change in atmospheric composition with height

The distributions of chemically stable gases, which have long residence times, such as O_2, N_2, and the inert gases, are determined by two competing physical processes: molecular diffusion and mixing due to macroscopic fluid motions. In the absence of sources, sinks, or turbulent mixing, diffusion by random molecular motions tends to produce an atmosphere in which the mean molecular weight of the mixture of gases gradually decreases with height to the point where only the lightest gases (hydrogen and helium) are present at the highest altitudes. The concentration $C(z)$ of a gas at height z above the Earth's surface is given by

$$C(z) = C(0)\exp\left(-\frac{z}{H}\right) \qquad (3.2)$$

where $C(0)$ is the concentration at $z = 0$ and H is the *scale height*[3] for the gas. It can be seen from Eq. (3.2) that if z is set successively equal to $H, 2H, 3H, \ldots$, then $C(z)/C(0)$ is equal to exp (–1), exp (–2), exp (–3), ... That is, the concentration of the gas decreases by a factor e (= 2.718) for each increase H in the height z above the Earth's surface. The scale height of a gas is inversely proportional to the molecular weight of the gas, therefore, the concentrations of the lighter gases decrease more slowly with height than those of the heavier gases.

In contrast to molecular diffusion, mixing due to the motions of macroscale air parcels does not discriminate on the basis of molecular weight. Within the range of heights in the atmosphere where mixing strongly predominates (i.e., in the homosphere), and provided chemistry does not play a dominant role, atmospheric composition tends to be independent of height (Fig. 3.1). The relative effectiveness of molecular diffusion increases in proportion to the root mean square velocity of the random molecular motions and the mean free path between molecular collisions. In mixing by fluid motions, the analog of the mean free path is the *mixing length*, which depends upon the spectrum of scales of motion present in the atmosphere.

Of the various factors that influence the relative effectiveness of molecular diffusion and mixing by fluid motions, by far the most important is the increase in the mean free path of molecules with height, which is illustrated in Figure 3.2. In the lower atmosphere the mean free path is

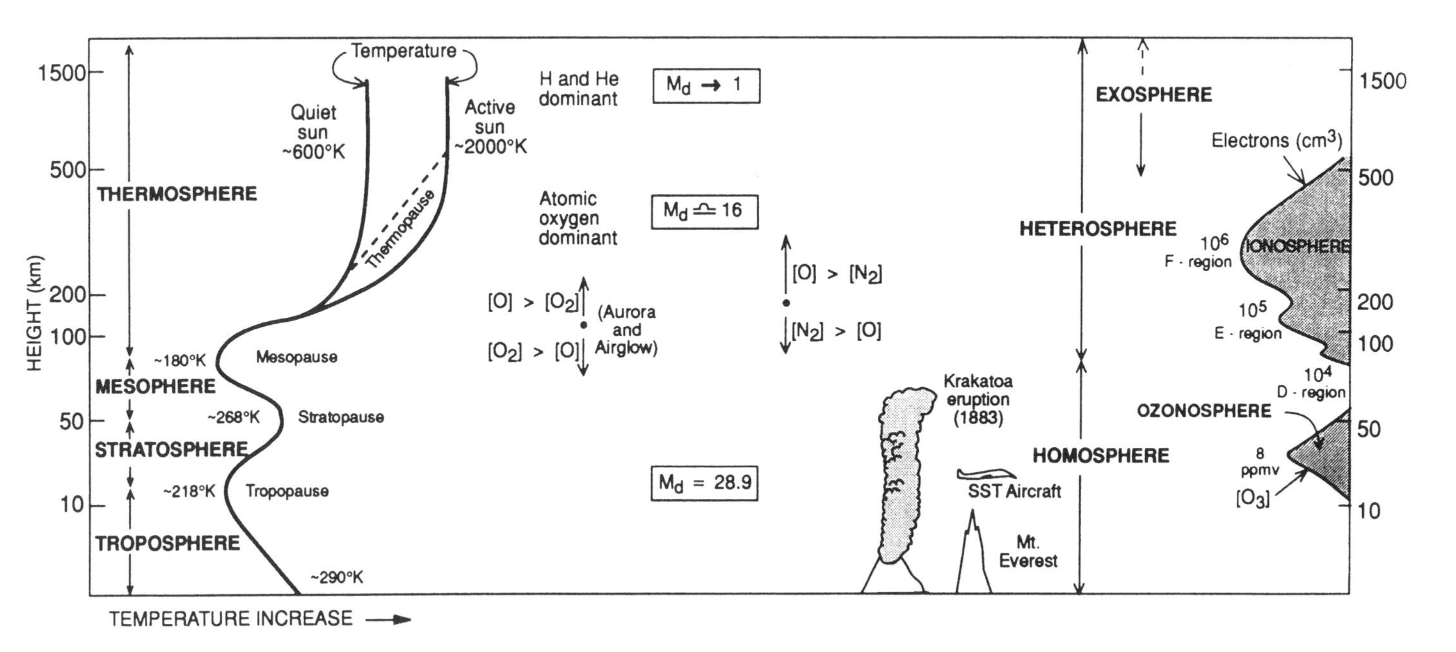

Figure 3.1. Two ways of dividing the atmosphere: by temperature structure (left side) and by composition (right side). The change in the apparent molecular weight of air (M_d) due to the changing composition of the atmosphere with height is shown in the center.

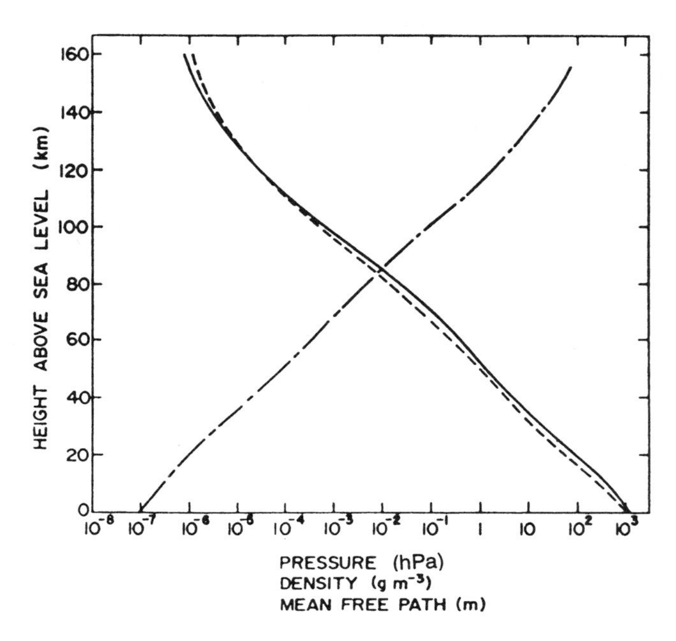

Figure 3.2. Vertical profiles of pressure (– – –), air density (——), and mean free molecular path (—- -—-) for the United States Extension to the International Civil Aviation Organization's Standard Atmosphere. [Adapted from *CRC Handbook of Chemistry and Physics*, 54th Edition, CRC Press, pp. F186–190 (1973).]

so short that the time required for the vertical separation of the heavier and lighter constituents by molecular diffusion is many orders of magnitude longer that the time required for turbulent fluid motions to homogenize them. Near an altitude of 100 km the two competing processes are of comparable importance, while well above 100 km the vertical mixing of atmospheric constituents is essentially controlled by molecular diffusion. The level of transition from turbulent mixing to molecular diffusion is called the *turbopause*. In the region above the turbopause, the atmosphere is not well mixed; this region is called the *heterosphere* (Fig. 3.1).

For reasons that will be discussed in Section 10.1, ozone concentrations rise rapidly above the tropopause, reach peak values between ~15 and 30 km, and then decline rapidly between ~30 and 50 km. This region is called the *ozonosphere* (Fig. 3.1).

The composition of the lower part of the heterosphere is strongly influenced by the photodissociation of diatomic oxygen, which gives rise to large numbers of free oxygen atoms. Above an altitude of about 120 km, most of the atmospheric oxygen is in the atomic form. It is in the lower part of the heterosphere, which contains relatively large concentrations of atomic oxygen and hydrogen as well as OH, NO, and O_3, that photochemical reactions produce products in excited electronic states, which emit weak but detectable luminescence known as *airglow*.[4]

Above an altitude of ~200 km there is a noticeable increase in the relative abundance of the lighter constituents, due to the effects of molecular mass on diffusion. The heaviest major constituent, diatomic nitrogen, drops off most rapidly with height. Around 500 km the atmosphere is predominantly atomic oxygen, with only traces of diatomic nitrogen and the very light constituents (He and H). Above 1,000 km, He and H are the dominant species.

The structure of the heterosphere is strongly dependent upon temperature, which varies by a factor of three or more in response to solar activity. At low temperatures the transition to lighter species takes place at relatively low levels, whereas, at high temperatures it takes place at higher levels. Above 300 km, the pressure and density at any level vary by an order of magnitude or more in response to changes in solar activity. Above ~500 km, the mean free path between molecular collisions is so long that individual molecules follow ballistic trajectories, like rockets. For all species of molecules there exists a single *escape velocity* (V_e) for which the kinetic energy of the molecule is equivalent to the potential energy that needs to be supplied to lift it out of the Earth's gravitational field. Escape velocity is a function of height only; in the Earth's atmosphere at a level of 500 km it is on the order of 11 km s^{-1}.

The most probable velocity of any molecular species is given by[5]

$$V_0 = \sqrt{\frac{2kT}{Mm_H}} \tag{3.3}$$

where, k is Boltzmann's constant, T the absolute temperature, M the molecular weight of the species, and m_H the mass of a hydrogen atom (1.67×10^{-27} kg). The individual molecules within a gas exhibit a distribution of velocities scattered about V_0. The kinetic theory of gases predicts that only about 2% of the molecules have velocities greater than $2V_0$, and only one molecule in 10^4 has a velocity greater than $3V_0$. Additional examples are given in Table 3.2.

In the Earth's atmosphere, temperatures at the base of the "escape region" (or *exosphere*, as it is called) are ~600 K (Fig. 3.1). If this value is substituted into Eq. (3.3) it is found that, for hydrogen atoms ($M = 1$),

Table 3.2. *Fraction of gas molecules with velocities V greater than various multiples of the most probable velocity V_0*

V/V_0	Fraction
1	0.5
2	0.02
3	10^{-4}
4	10^{-6}
6	10^{-20}
10	10^{-50}
15	10^{-90}

$V_0 \simeq 3\,\mathrm{km\,s^{-1}}$. Therefore, according to Table 3.2, for each collision near 500 km, the probability of escape (that is $V > V_e$) is slightly greater than 10^{-6}. The corresponding time period required for the escape of all the hydrogen from the Earth's atmosphere turns out to be much less that the lifetime of the Earth. This is one reason for the relative absence of free hydrogen in the atmosphere, despite its continual production due to the dissociation of water (see Exercises 2 and 3 in Appendix I). For atomic oxygen ($M = 16$), $V_0 \simeq 0.8\,\mathrm{km\,s^{-1}}$, and the probability of escape is $\simeq 10^{-84}$. The rate of escape of atomic oxygen is so slow that the cumulative loss over the lifetime of the Earth is negligible.

Because of the rapid increase in the mean free path of particles with height, the decrease in pressure with height, and the transition to more stable species of ions at higher levels, the free electrons produced by the Sun's ionizing radiation in the upper atmosphere have much longer lifetimes that those resulting from the various sources at lower levels. Hence, most of the free electrons in the atmosphere are located above 60 km, where they exist in sufficient numbers to affect radio wave propagation. This region of the atmosphere is called the *ionosphere* (Fig. 3.1).

A vertical profile of the number density of free electrons is shown in Figure 3.1. It can be seen that the concentration of free electrons increases with height from very small values below ~60 km to a maximum value near 300 km. Some small undulations, labeled D and E, are shown in Figure 3.1. These irregularities in the vertical gradient of electron concentration have a profound effect upon the propagation of radio waves. Before the results shown in Figure 3.1 were established (by means of *in situ* measurements from rockets and satellites) it was widely believed, on the basis of radio wave propagation experiments, that the bump at E (~110km) corresponded to a distinct maximum in electron density. At that time, the terms *E-layer* and *F-layer* came into use as a means of distinguishing between the lower and upper maxima. The term *D-region* was used to denote a lower layer in which strong absorption of radio waves takes place due to collisions between electrons and neutral particles. Later it was discovered that the D-region contains a separate bump in the electron density profile.

Electron densities in the ionosphere decline after sunset, in response to the interruption of the ionizing flux of solar radiation. The extent of the decrease varies with height. Most of the electrons in the D- and E-regions recombine with positive ions during the night, and some electrons in the D-region undergo attachment to form negative ions. With the virtual disappearance of the D-region, the absorption of radio waves is reduced to the point where signals reflected from higher layers of the ionosphere interfere with reception in the AM band of radios. The F-region exhibits a smaller diurnal variability, which is often masked by irregular fluctuations in electron density.

Electron density also varies in response to events on the Sun that moderate x-rays (one form of short-wavelength ionizing radiation) reaching the ionosphere. For example, strong solar flares are accompanied by bursts of x-rays that increase the electron densities within the D-region on the daylight side of the Earth. Enhanced electron densities give rise to increased absorption of radio waves, which sometimes causes fade-outs in long-range communications that depend upon the reflection of radio waves from the ionosphere.

Within the E-region, the mean free path is sufficiently short that the movement of positive ions is largely controlled by the drift of the neutral constituents, which account for an overwhelming fraction of the mass. However, within this same region, the free electrons are constrained to move along the magnetic field lines. Hence, whenever the neutral atmosphere within the E-region moves across the lines of the Earth's magnetic field, charge separation occurs, currents flow, and voltages are induced; the effect is analogous to the generation of electrical power in a dynamo. Currents generated in the E-region are responsible for variations in geomagnetism and in the structure of the F-region. An important input of energy into the "dynamo" is associated with atmospheric tidal motions driven not by gravitational effects but by the diurnal variation in solar heating.

In this chapter we have ventured into the realms of *aeronomy*, which is concerned with the atmosphere above about 50 km. We must now return to the main subject of this book, which is the chemistry of the troposphere and the stratosphere.

Exercises

See Exercises 1(h) and (i), and Exercises 10–13 in Appendix I.

Notes

1 Ludwig Boltzmann (1844–1900). Austrian physicist. Made fundamental contributions to the kinetic theory of gases. Committed suicide.
2 A *radical*, or *free radical*, is an atom or molecule containing an unpaired electron. Radicals are usually very reactive and therefore short-lived.
3 For a discussion of scale height see *Atmospheric Science: An Introductory Survey* by J. M. Wallace and P. V. Hobbs, Academic Press, New York, pp. 48–56 (1977).
4 Airglow should not be confused with the *aurora*, which occurs in the same general region of the atmosphere (although generally confined to polar regions). The aurora is a more intense emission that is produced by the bombardment of the upper atmosphere by electrons and protons from the Sun.
5 For a derivation of Eq. (3.3) see, for example, *Physics for Scientists and Engineers* by R. A. Serway, Saunders College Publishing, Orlando, Florida, 3rd ed., pp. 574–575 (1990).

4

Interactions of solar and terrestrial radiation with atmospheric trace gases and aerosols

This chapter is concerned with the attenuation by atmospheric gases and aerosols of the incoming shortwave radiation from the Sun (*solar radiation*), and the emission and absorption of outgoing longwave radiation (also called *infrared*, *terrestrial*, or *thermal radiation*) from the Earth and its atmosphere. These interactions play important roles in determining the energy balances, and therefore the temperatures, of the Earth's surface and the atmosphere. Also, the absorption of solar radiation by some atmospheric constituents can lead to photochemical reactions, which play crucial roles in atmospheric chemistry.

Figure 4.1 shows current best estimates of the annual global energy balance of the Earth-atmosphere system expressed in terms of 100 units of incoming solar radiation at the top of the Earth's atmosphere (TOA). About 22 of these units are reflected back into space by clouds, aerosols, and gases, about 20 units are absorbed by the atmosphere, 9 units are reflected from the Earth's surface, and the remaining 49 units are absorbed at the Earth's surface. Measured in the same units, the Earth receives an additional 95 units due to longwave radiation from the atmosphere. Therefore, the total energy received by the Earth is 144 units. Thermal equilibrium at the surface of the Earth is achieved by these 144 units being transferred back to the atmosphere: 114 of them are radiated to the atmosphere as longwave radiation, 23 units as evapotranspiration (which is ultimately released to the atmosphere as latent heat in precipitation), and 7 units are transferred to the atmosphere by heat fluxes associated with turbulence, convection, and so on.

The atmosphere itself must also be in thermal equilibrium. In terms of our units the atmosphere absorbs about 20 from solar radiation and 102 from longwave radiation from the Earth's surface (the other 12 units pass through the atmosphere unattenuated through the so-

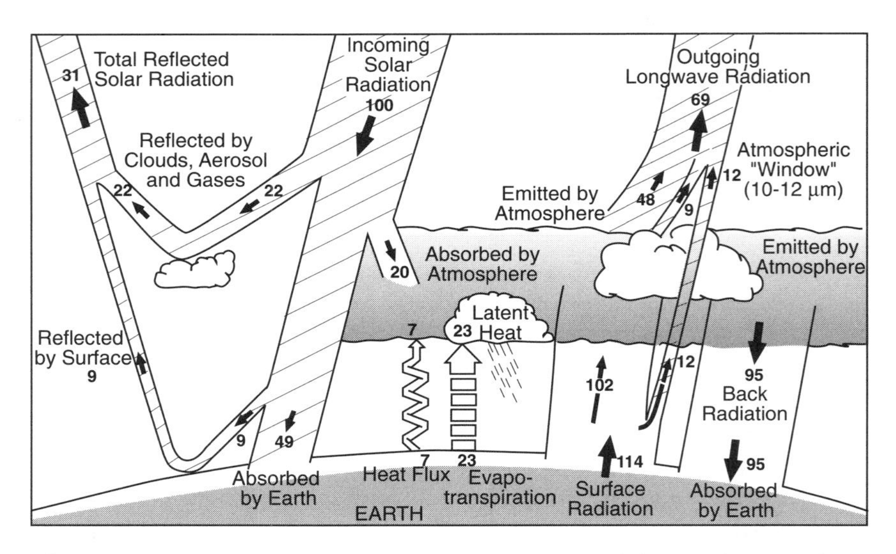

Figure 4.1. The annual mean global energy balance of the Earth-atmosphere system. Numbers are given as percentages of the globally averaged solar energy incident upon the top of the atmosphere (TOA). The 100 units of incoming solar radiation at the TOA represents $342\,W\,m^{-2}$ (see text).

called infrared window), and 30 from latent heat and thermals, for a total receipt of 152 units. The atmosphere "loses" 95 units through longwave radiation to the Earth's surface and 57 to space from longwave radiation from gases, aerosols, and clouds. Note that at the TOA 100 units of energy are incoming and 100 (69 from outgoing longwave radiation and 31 from reflected solar radiation) are outgoing (Figure 4.1), which ensures thermal equilibrium for the Earth-atmosphere system as a whole.

Note that whereas the disposition of solar radiation is dominated by absorption and scattering, thermal radiation is dominated by emission and absorption.

4.1 Some basic concepts and definitions[1]

The rate of transfer of energy by electromagnetic (em) radiation is called *radiant flux* (units are joules per second, $J\,s^{-1}$, or watts, W). The radiant flux incident on a unit area is called the *irradiance* ($W\,m^{-2}$), denoted by E. The irradiance per unit wavelength interval, centered on wavelength λ, is denoted by E_λ ($W\,m^{-2}\,\mu m^{-1}$).

Exercise 4.1. (a) The radiant flux from the Sun is 3.90×10^{26} W. What is the irradiance at the outermost visible layer of the Sun, which is located at a distance of 7×10^8 m from the center of the Sun?

(b) Calculate the *equivalent blackbody temperature* of the outermost visible layer of the Sun.

(c) If the average irradiance from the Sun incident on a surface perpendicular to the direction of propagation of the solar beam at the TOA is 1,368 W m^{-2} (called the *solar constant*), what is the solar energy (in W m^{-2}) at the TOA when averaged over the whole surface of the globe? (Ignore the thickness of the atmosphere compared to the radius of the Earth.)

Solution. (a) Irradiance is the radiant flux passing through 1 m^2. Since the Sun's radiation can be considered to propagate radially outward in all directions from the center of the Sun, it will everywhere be normal to a sphere centered on the Sun. A sphere of radius 7×10^8 m has a surface area of $4\pi (7 \times 10^8)^2$ m^2. Therefore, the irradiance passing through this sphere is $3.90 \times 10^{26}/4\pi (7 \times 10^8)^2 = 6.34 \times 10^7$ W m^{-2}.

(b) The equivalent blackbody temperature (T_E) of an object is the temperature that a blackbody (i.e., a perfect radiator and a perfect absorber) would have to have in order to emit the same amount of radiation as the object. The total irradiance (over all wavelengths) from a blackbody is given by the *Stefan*[2]*– Boltzmann law*

$$E^* = \sigma T^4 \tag{4.1}$$

where σ is the Stefan–Boltzmann constant, which has a value of 5.67×10^{-8} W m^{-2} deg^{-4}, and T is the temperature of the blackbody (in K). When T in Eq. (4.1) is equal to T_E for the Sun's outermost visible layer, E^* given by Eq. (4.1) is equal to the irradiance of the outermost visible layer of the Sun calculated in (a). Therefore,

$$\left(5.67 \times 10^{-8}\right) T_E^4 = 6.34 \times 10^7$$

or

$$T_E = 5{,}780\,\mathrm{K}$$

(c) If the solar irradiance at the TOA is 1,368 W m^{-2}, 1,368 W is incident on every square meter of surface that is oriented perpendicular to the Sun's rays at the TOA. If the Earth's radius is R_E, the hemisphere of the Earth that faces the Sun at any given time has an area πR_E^2 when projected onto a plane perpendicular to the Sun's rays. (Note: Because of its large distance from the Earth, the Sun's rays can be assumed to be parallel at the TOA.) Therefore, the total radiant flux incident on the Earth is 1,368 πR_E^2. However, the total surface area of the Earth is $4\pi R_E^2$. Therefore, when averaged over the whole surface area of the Earth, the average solar energy at the TOA is 1,368 $\pi R_E^2/4\pi R_E^2$ or 342 W m^{-2}. (This means that in Figure 4.1, the 100 units of incoming solar radiation at the TOA represents 342 W m^{-2}. Therefore, the various percentages given in Figure 4.1 must be multiplied by 342 to convert them to W m^{-2}. For example, averaged over the globe the solar energy absorbed by the atmosphere is 20% of 342, or about 68 W m^{-2}.)

The wavelength (λ_m) of peak radiant flux from a blackbody is given by the *Wien*[3] *displacement law*

$$\lambda_m = \frac{2,897}{T} \tag{4.2}$$

where T is the temperature (in K) of the blackbody and λ_m is in μm. Application of this law shows that the Sun (with its high value of T) emits its peak radiant flux at a wavelength of 0.5 μm, while the radiant flux emitted by the Earth and its atmosphere (which have much lower values of T) is largely confined to infrared (IR) radiation (Figure 4.2).

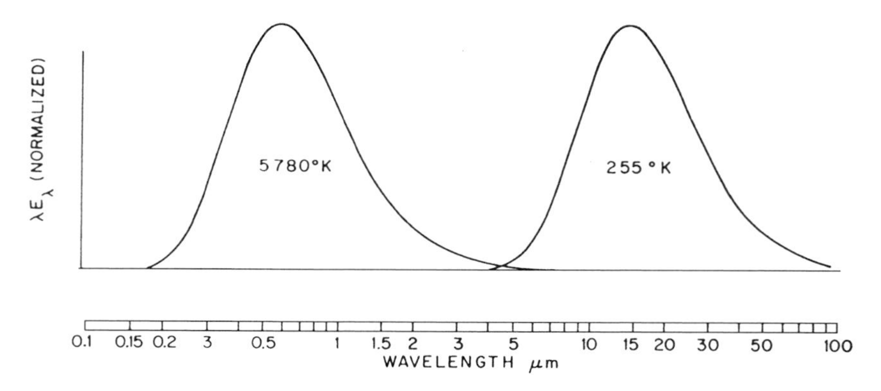

Figure 4.2. Normalized blackbody spectra representative of the Sun (left) and the Earth (right). The ordinate is multiplied by wavelength to make the area under each of the two curves proportional to irradiance, then each curve has been normalized.

The *emissivity* (ε_λ) at wavelength λ of a body is defined as

$$\varepsilon_\lambda \equiv \frac{E_\lambda}{E^*_\lambda} \tag{4.3}$$

where E_λ is the irradiance at wavelength λ from the body at temperature T and E^*_λ the irradiance from a blackbody at T. The *absorptivity* a_λ of a body is defined in a similar way, namely, the ratio of the irradiance at wavelength λ absorbed by a body to the irradiance incident upon it (all of which is absorbed by a blackbody).

Materials that are strong absorbers of em radiation at particular wavelengths are strong emitters at those same wavelengths (because at these wavelength the molecules of the material readily vibrate or rotate, so they both absorb and emit radiation efficiently). This relationship is summarized by *Kirchhoff's*[4] *law*

$$a_\lambda = \varepsilon_\lambda \tag{4.4}$$

Consider a parallel beam of solar radiation propagating vertically downward through the atmosphere. Let the irradiances at heights z and z + dz above the Earth's surface be $E_\lambda(z)$ and $E_\lambda(z) + dE_\lambda(z)$, respectively, and the irradiance at the TOA be $E_{\lambda\infty}$ (Fig. 4.3). Atmospheric gases and aerosols can remove energy from the beam by scattering and absorbing radiation. It can be shown (both experimentally and theoretically) that a thin section dz of air scatters and absorbs em radiation of wavelength λ in an amount proportional to dz and to $E_\lambda(z)$. Therefore, we can write

$$dE_\lambda(z) = (b_{s\lambda} + b_{a\lambda})E_\lambda(z)\,dz \tag{4.5}$$

or

$$dE_\lambda(z) = b_{e\lambda}E_\lambda(z)\,dz \tag{4.6}$$

where

$$b_{e\lambda} \equiv b_{s\lambda} + b_{a\lambda} \tag{4.7}$$

and $b_{s\lambda}$, $b_{a\lambda}$, and $b_{e\lambda}$ are called the *scattering*, *absorption*, and *extinction coefficients* of the air, respectively. From dimensional considerations, it can be seen from Eqs. (4.5) and (4.6) that $b_{s\lambda}$, $b_{a\lambda}$, and $b_{e\lambda}$ have dimensions of inverse meters (m^{-1}). Note that since z is height above the Earth's surface, dz in Figure 4.3 is negative as is $dE_\lambda(z)$.

Scattering, absorption, and extinction coefficients can be defined for gases and aerosols individually. For aerosols, $b_{s\lambda}$, $b_{a\lambda}$, and $b_{e\lambda}$ vary greatly

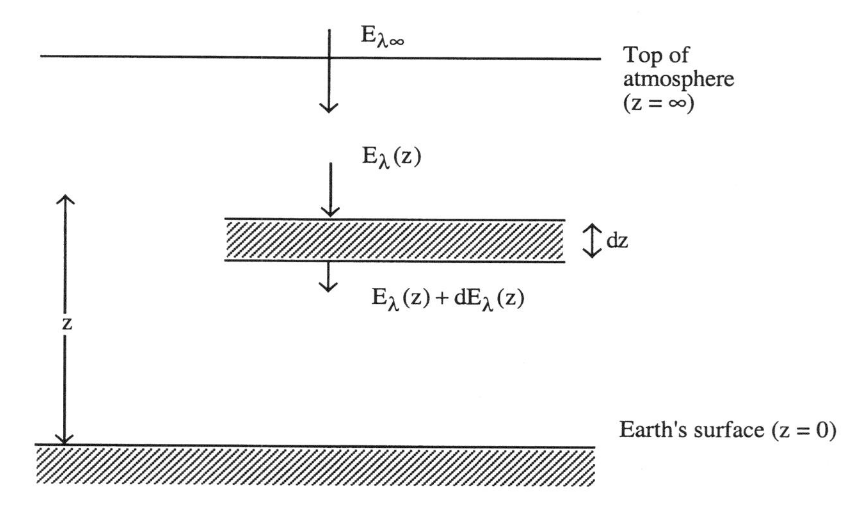

Figure 4.3. Schematic of a parallel beam of solar radiation propagating vertically downward through the atmosphere.

both with the concentration and the size of the particles (relative to λ). For highly polluted urban air, the values of $b_{s\lambda}$ near the surface and at mid-visible wavelengths are on the order $2 \times 10^{-4}\,m^{-1}$. For clean rural air, $b_{s\lambda}$ is $\sim 3 \times 10^{-5}\,m^{-1}$. For aerosol-free air at sea level, $b_{s\lambda}$ at midvisible wavelengths is $\sim 1.3 \times 10^{-5}\,m^{-1}$ (due to scattering by gaseous molecules).

The parameters $b_{s\lambda}$, $b_{a\lambda}$, and $b_{e\lambda}$ can be converted to *mass scattering*, *absorption*, and *extinction efficiencies* (represented by the symbol α with an appropriate subscript) by dividing them by the mass concentration of the gas (or aerosol) being considered. For example, the absorption mass efficiency of air at wavelength λ is given by

$$\alpha_{a\lambda} = b_{a\lambda}/\rho \tag{4.8}$$

where $b_{a\lambda}$ is the absorption coefficient for air, and ρ the density of the constituent. Since $b_{a\lambda}$ has units of m^{-1} and ρ has units $kg\,m^{-3}$, the units of $\alpha_{a\lambda}$ are $m^2\,kg^{-1}$. The most important absorbing aerosol for solar radiation is *black carbon*, which has an absorption mass efficiency of about $10\,m^2\,g^{-1}$.

Integration of Eq. (4.6) from height z above the Earth's surface to the TOA, where the irradiance is the solar spectral irradiance, $E_{\lambda\infty}$, gives

$$\int_{E_\lambda(z)}^{E_{\lambda\infty}} \frac{dE_\lambda(z)}{E_\lambda(z)} = \int_z^\infty b_{e\lambda}\,dz \tag{4.9}$$

The *optical depth* at wavelength λ of the atmosphere between height z and the TOA is defined as

$$\tau_\lambda \equiv \int_z^\infty b_{e\lambda} dz \tag{4.10}$$

(Note that, despite its name, τ_λ is a dimensionless quantity since $b_{e\lambda}$ has units of m^{-1}.) From Eqs. (4.9) and (4.10)

$$\ln E_{\lambda\infty} - \ln E_\lambda(z) = \tau_\lambda$$

or

$$E_\lambda(z) = E_{\lambda\infty} \exp(-\tau_\lambda) \tag{4.11}$$

Equation (4.11) is called *Beer's*[5] *law*.

Exercise 4.2. If the *total column optical depth of the atmosphere* (i.e., the optical depth from the Earth's surface to the TOA) at mid-visible wavelengths is 0.4, what is the percentage reduction in solar irradiance between the TOA and sea level for a vertical Sun?

Solution. The percentage reduction is solar irradiance between the TOA (E_∞) and sea level is

$$\frac{E_\infty - E_{\text{sea level}}}{E_\infty} 100 \tag{4.12}$$

From Eq. (4.11)

$$E_{\text{sea level}} = E_\infty \exp(-\tau_\lambda) \tag{4.13}$$

From Eqs. (4.12) and (4.13), the percentage reduction in solar irradiance can be written as

$$[1 - \exp(-\tau_\lambda)]100$$

If, at midvisible wavelengths, $\tau = 0.4$, the percentage reduction in solar irradiance is

$$[1 - \exp(-0.4)]100 = 33\%$$

This reduction of about one-third in the solar intensity at mid-visible wavelengths in a polluted atmosphere is due primarily to aerosols.

We see from Eq. (4.11) that if $\tau_\lambda = 1$ the irradiance at height z above the Earth's surface would be a factor exp (−1), or about one-third, of the irradiances at the TOA. Typical values of the total column optical depth of the atmosphere at midvisible wavelengths are ~0.3 to 0.5 for urban air and ~0.2 for fairly clean air.

The *transmissivity* (or *transmittance*) of the atmosphere lying above height z is defined as

$$T_\lambda \equiv \frac{E_\lambda(z)}{E_{\lambda\infty}} = \exp(-\tau_\lambda) \tag{4.14}$$

The em energy that is scattered and absorbed between the TOA and height z is $1 - T_\lambda$. In the absence of scattering, the absorptivity (defined as the fraction of $E_{\lambda\infty}$ that is absorbed between the TOA and height z) is

$$a_\lambda = 1 - T_\lambda = 1 - \exp(-\tau_\lambda) \tag{4.15}$$

Equation (4.15) shows that a_λ approaches unity exponentially as the optical depth τ_λ increases.

From Eq. (4.10), the optical depth of a layer of the atmosphere lying between heights z_1 and z_2 ($z_2 > z_1$) is

$$\tau_\lambda = \int_{z_1}^{z_2} b_{e\lambda} dz$$

If $b_{a\lambda} \gg b_{s\lambda}$, and if $b_{a\lambda}$ is independent of z between z_1 and z_2,

$$\tau_\lambda \simeq b_{a\lambda}(z_2 - z_1)$$

where $z_2 - z_1$ is the *path length* of the layer. At wavelengths close to the center of strong gaseous absorption lines $b_{a\lambda}$ is large, so that very small path lengths can produce large values of τ_λ, which absorb virtually all of the incident radiation. By contrast, at wavelengths away from absorption lines, very long path lengths are required to produce appreciable absorption.

The nonlinear relationship between a_λ and τ_λ, given by Eq. (4.15), causes individual lines in the absorption spectrum of a gas (i.e., a plot of a_λ versus λ) to broaden and merge into *absorption bands* as τ_λ increases. For sufficiently short path lengths $\tau_\lambda \ll 1$, and Eq. (4.15) becomes

$$a_\lambda = 1 - T_\lambda \simeq 1 - (1 - \tau_\lambda) = \tau_\lambda \simeq b_{a\lambda}(z_2 - z_1)$$

In this case, there is a linear relationship between a_λ and the path length. As the path length increases the relationship between a_λ and $(z_2 - z_1)$ becomes nonlinear, with a_λ asymptotically approaching unity (i.e., complete absorption) over an ever-widening span of wavelengths. This is depicted in Figure 4.4. As the individual absorption lines widen, adjacent

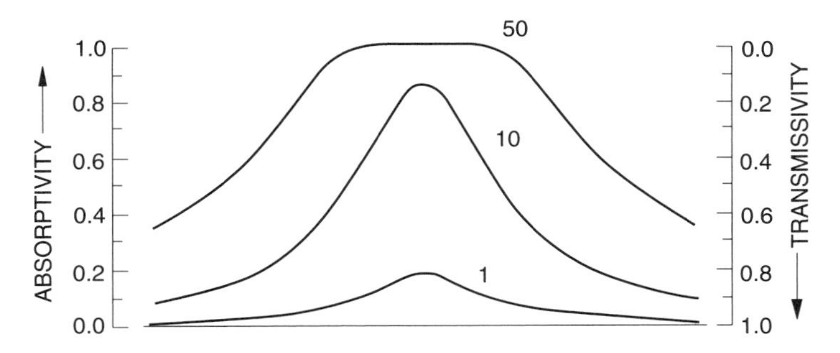

Figure 4.4. Absorption spectrum of an idealized spectral line for three path lengths with relative values indicated on the lines.

lines begin to progressively overlap. Consequently, for a certain range of values of the path length (or optical depth), line clusters become manifested as *absorption bands*, in which substantial absorption occurs over a range of wavelengths. The quasi-transparent regions that lie between absorption bands are called *windows*.

4.2 Attenuation of solar radiation by gases

Figure 4.5 shows the spectrum of solar radiation at the TOA (thick upper curve) and at sea level (thick lower curve) for a hypothetical aerosol-free atmosphere. In the absence of aerosols, the reduction in solar irradiance between the top and the bottom of the atmosphere is due to scattering and absorption by gaseous molecules. The contribution of absorption to the attenuation of the solar radiation is indicated by the shaded region in Figure 4.5. Therefore, in the absence of absorption by gases (and attenuation by clouds and aerosols), the spectrum of solar radiation at sea level would be given by the thin curve in Figure 4.5.

The scattering of visible light by gaseous molecules (which have dimensions much smaller than the wavelength of visible light) is called *molecular* or *Rayleigh*[6] *scattering*. In this case, for a molecule with a given refractive index,[7] the fraction of light that is scattered is proportional to $(r/\lambda)^4$, where r is the radius of the molecule and λ the wavelength of the em radiation. For example, the relative amounts of em scattering of blue light ($\lambda \simeq 0.47\,\mu$m) and red light ($\lambda \simeq 0.64\,\mu$m) by a molecule of a given size is

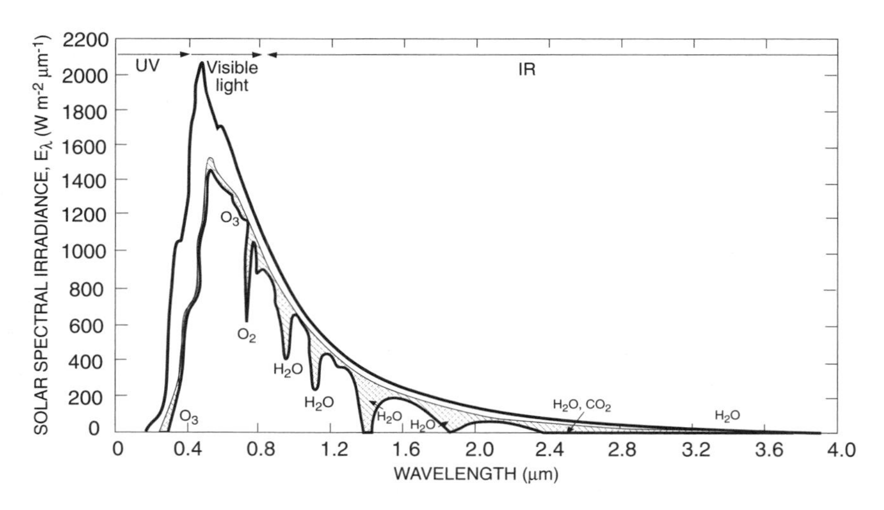

Figure 4.5. The upper thick curve shows the solar irradiance at the top of the atmosphere, and the lower thick curve the computed solar irradiance at sea level for the hypothetical case of an aerosol-free atmosphere. The thin curve represents the irradiance at sea-level if scattering by gaseous molecules alone attenuated solar radiation. The shaded area shows the contribution to the reduction in solar irradiance due to absorption by atmospheric gases, with the main gaseous absorbers indicated. [Adapted from *An Introduction to Atmospheric Radiation* by K.-N. Liou, Academic Press (1980).]

$$\frac{\text{scattering by blue light}}{\text{scattering by red light}} = \left(\frac{\lambda_{\text{red}}}{\lambda_{\text{blue}}}\right)^4 = \left(\frac{0.64}{0.47}\right)^4 = 3.5$$

The blueness of the sky is a consequence of the much greater amount of blue light, compared to red light, scattered by molecules in the air. The total column optical depth of the atmosphere due to Rayleigh scattering alone is about 0.1 at midvisible wavelengths. Rayleigh scattering is proportional to the density of the air, and therefore decreases with increasing altitude.

For Rayleigh scattering the radiation is evenly distributed between the forward and backward directions. For example, averaged over different Sun angles and wavelengths, atmospheric gases scatter about 6% of the incoming solar radiation back to space and about 6% reaches the Earth's surface as *diffuse* radiation.

In contrast to scattering, the absorption of solar radiation by atmospheric gases is quite variable with wavelength (the shaded areas in Fig. 4.5). This is because molecules can absorb em radiation by increasing their *translational* kinetic energy (which increases the temperature of the gas), the *vibrational* energy of their atoms, the *rotational* energy of the molecules about their centers of mass, and their *electronic* energy. Changes in translational energy are not relevant in the troposphere or stratosphere. Changes in vibrational, rotational, and electronic energies are important, and these can change only by discrete amounts (i.e., they are *quantitized*).

Shown in Figure 4.5 are the main atmospheric gaseous absorbers of solar radiation from the near ultraviolet (UV), through visible wavelengths, to the near IR region of the em spectrum. The UV radiation from the Sun at $\lambda < 0.3\,\mu m$ is absorbed in the upper atmosphere, where it provides the main source of energy for the dynamics of that region. The UV radiation between 0.3 and $0.4\,\mu m$ reaches the Earth's surface.

In the near-UV ($\lambda \simeq 0.2$–$0.3\,\mu m$), the absorption of solar radiation is due mainly to electronic transitions of O_3 in the stratosphere, which will be discussed in Chapter 10. At shorter wavelengths in the UV ($\lambda < 0.2\,\mu m$), O_2 and N_2 also absorb solar radiation (not shown in Fig. 4.5).

There is relatively little absorption of solar radiation in the visible region (Fig. 4.5). Ozone absorbs weakly in the visible and near IR regions from about 0.44 to $1.18\,\mu m$. Molecular oxygen has two weak absorption bands at red wavelengths. The O_2 band at $0.7\,\mu m$ is important because of the large solar irradiance at this wavelength[8]; this band led to the discovery of the isotopes oxygen-18 and oxygen-17.

Water vapor has several important absorption bands in the near IR centered at 0.94, 1.1, 1.38, and $1.87\,\mu m$ (Fig. 4.5), which are due to vibrational and rotational energy transitions. There are also water vapor bands at 2.7 and $3.2\,\mu m$. Carbon dioxide has very weak absorption bands at 1.4, 1.6, and $2.0\,\mu m$ (not resolved in Fig. 4.5); it also has a band at $2.7\,\mu m$, which overlaps the water vapor band at this wavelength.

4.3 Vertical profile of absorption of solar radiation in the atmosphere[9]

We can illustrate some of the basic principles involved in the absorption of solar radiation as it passes through the atmosphere by considering a dry atmosphere in which the temperature does not vary with height (a so-called *isothermal atmosphere*), and by considering the case of an overhead Sun. Since we are interested in absorption, rather than scattering, we will assume that $b_{e\lambda} = b_{a\lambda}$. We will also assume that $\alpha_{a\lambda}$ does not vary with height, which could be the case, for example, for absorption in a CO_2 band because CO_2 is well mixed in the atmosphere. (Note that since $b_{a\lambda} = \rho\alpha_{a\lambda}$, $b_{a\lambda}$ varies with height even when $\alpha_{a\lambda}$ is constant.)

For an isothermal atmosphere at temperature T, the density of the air at height z above the Earth's surface is given by

$$\rho(z) = \rho_0 \exp\left(-\frac{z}{H}\right) \tag{4.16}$$

where ρ_0 is the air density at the surface, and H is the scale height for air density which is given by

$$H = \frac{R_d T}{g} \tag{4.17}$$

Because we are assuming that the absorber is well mixed, its scale height is the same as that of the air. R_d is the gas constant for 1 kg of dry air, and g is the acceleration due to gravity (compare Eq. (4.16) with Eq. (3.2)). From Eqs. (4.8), (4.10), and (4.16), and using the assumptions stated earlier,

$$\tau_\lambda = \int_z^\infty b_{e\lambda} dz = \int_z^\infty b_{a\lambda} dz = \int_z^\infty \rho\alpha_{a\lambda} dz = \alpha_{a\lambda}\rho_0 \int_z^\infty \exp\left(-\frac{z}{H}\right) dz$$

or, carrying out the integration,

$$\tau_\lambda = H\alpha_{a\lambda}\rho_0 \exp\left(-\frac{z}{H}\right) \tag{4.18}$$

The incident radiation absorbed within any differential layer of the atmosphere is

$$-dE_\lambda = E_{\lambda\infty} T_\lambda da_\lambda \tag{4.19}$$

where T_λ is the transmissivity of the portion of the atmosphere that lies above the layer being considered. From the definition of a_λ and applying Eq. (4.6) for extinction due to absorption alone

$$da_\lambda \equiv -\frac{dE_\lambda}{E_\lambda} = -b_{a\lambda} dz = -\rho\alpha_{a\lambda} dz \tag{4.20}$$

Substituting for T_λ from Eq. (4.14) and da_λ from Eq. (4.20) into Eq. (4.19) yields

$$dE_\lambda = E_{\lambda\infty} \exp(-\tau_\lambda)\rho\alpha_{a\lambda} dz$$

Using Eq. (4.16) in this last expression

$$dE_\lambda = E_{\lambda\infty}\alpha_{a\lambda}\rho_0 \exp\left(-\frac{z}{H}\right)\exp(-\tau_\lambda)dz$$

Substituting for exp (–*z*/*H*) from Eq. (4.18), we obtain an expression for the absorption per unit thickness of the layer as a function of optical depth

$$\frac{dE_\lambda}{dz} = \frac{E_{\lambda\infty}}{H}\tau_\lambda \exp(-\tau_\lambda) \tag{4.21}$$

Equation (4.21) is important because it represents the energy absorbed per unit volume of air, per unit time, and per unit wavelength.

The general shape of the curve dE_λ/dz versus z can be predicted as follows. If $\alpha_{a\lambda}$ is constant with z, we see from Eqs. (4.8) and (4.20) that

$$\frac{dE_\lambda}{dz} \propto E_\lambda \rho$$

High in the atmosphere E_λ is large, but the density of the air ρ is small; low in the atmosphere E_λ is small, but ρ is large. Hence, dE_λ/dz should have a peak value at some intermediate height in the atmosphere. At the level in the atmosphere where dE_λ/dz has a maximum value, $(d/dz)\,(dE_\lambda/dz) = 0$. From Eq. (4.21) this level is given by

$$\frac{E_{\lambda\infty}}{H}\frac{d}{dz}[\tau_\lambda \exp(-\tau_\lambda)] = 0$$

which yields $\tau_\lambda = 1$. That is, the *strongest absorption per unit thickness occurs at a level in the atmosphere where the slant path optical depth is unity*. A schematic of the variation of dE_λ/dz, E_λ, and ρ with height z is shown in Figure 4.6.

As we have seen, except in a very highly polluted atmosphere, the optical depth at visible wavelengths does not reach a value of unity even at the Earth's surface. However, at UV wavelengths, strong absorption by ozone results in an optical depth of unity in the stratosphere.

4.4 Heating of the atmosphere due to gaseous absorption of solar radiation

The absorption of solar radiation by various trace gases results in heating of the air. The rate of this heating can be computed as follows.

Let the downward directed irradiances at wavelength λ at heights $z + dz$ and z above the Earth's surface be $E_\lambda^\downarrow(z + dz)$ and $E_\lambda^\downarrow(z)$, respectively,

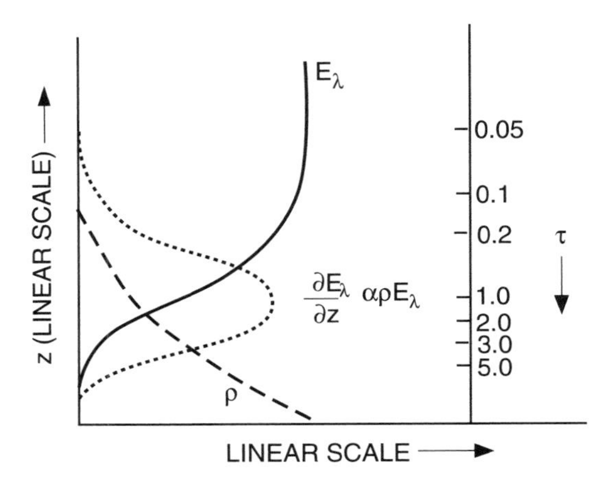

Figure 4.6. Schematic of the variations of E_λ, ρ, and dE_λ/dz with height z in an isothermal and well mixed atmosphere.

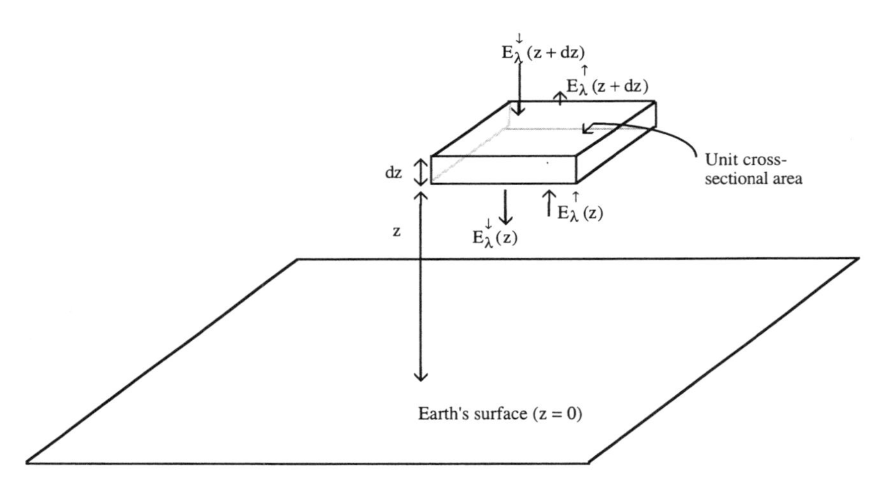

Figure 4.7. Flux divergence of radiation through an infinitesimal layer of air of thickness dz and unit cross-sectional area located at height z above the Earth's surface.

and the upward-directed irradiances at heights $z + dz$ and z be $E_\lambda^{\uparrow}(z + dz)$ and $E_\lambda^{\uparrow}(z)$, respectively (Fig. 4.7). Then the *net* irradiance in the downward direction at height $z + dz$ is

$$E_\lambda(z+dz) = E_\lambda^{\downarrow}(z+dz) - E_\lambda^{\uparrow}(z+dz) \tag{4.22}$$

and the net irradiance in the downward direction at height z is

$$E_\lambda(z) = E_\lambda^\downarrow(z) - E_\lambda^\uparrow(z) \tag{4.23}$$

The difference between the net downward-directed irradiances at the bottom and at the top of the layer of thickness dz is

$$dE_\lambda(z) = E_\lambda(z) - E_\lambda(z+dz) \tag{4.24}$$

which is called the *flux divergence* at height z. If radiation is absorbed in the layer dz, the flux divergence $dE_\lambda(z)$, defined by Eq. (4.24), will be negative.

If we assume that there are no losses of radiant energy through the vertical walls of the layer and there is no upward scattering, it follows from the definition of the absorptivity a_λ of the layer dz that

$$dE_\lambda(z) = -E_\lambda^\downarrow(z+dz)a_\lambda \tag{4.25}$$

Since the absorbed energy given by Eq. (4.25) heats the layer of air, we can write

$$dE_\lambda(z) = -\rho\, dz\, c_p \frac{dT}{dt} \tag{4.26}$$

where ρ is now the density of the air in the layer of thickness dz (and therefore ρdz is the mass of a unit horizontal area of this layer), c_p the specific heat at constant pressure of the air (units: J kg^{-1} deg^{-1}), and dT/dt the rate of temperature rise of the layer of air due to the absorption of em radiation. From Eqs. (4.25) and (4.26)

$$\frac{dT}{dt} = -\frac{1}{c_p\rho}\frac{dE_\lambda(z)}{dz} = \frac{a_\lambda}{c_p\rho}\frac{E_\lambda^\downarrow(z+dz)}{dz} \tag{4.27}$$

Exercise 4.3. Express dT/dt in terms of $dE_\lambda(p)$, dp, and the *dry adiabatic temperate lapse rate* Γ_d [$\equiv -(dT/dz)$ for dry parcels of air moving around in the atmosphere under adiabatic conditions], where p is the atmospheric pressure at height z above the Earth's surface.

Solution. The rate at which pressure changes with height in the atmosphere is given by the *hydrostatic equation*[10]

$$\frac{dp}{dz} = -g\rho \tag{4.28}$$

where g is the acceleration due to gravity, and ρ the density of the air at height z. From Eqs. (4.27) and (4.28),

$$\frac{dT}{dt} = \frac{g}{c_p} \frac{dE_\lambda(p)}{dp} \tag{4.29}$$

The dry adiabatic lapse rate is given by

$$\Gamma_d = \frac{g}{c_p} \tag{4.30}$$

Therefore, from Eqs. (4.29) and (4.30),

$$\frac{dT}{dt} = \Gamma_d \frac{dE_\lambda(p)}{dp} \tag{4.31}$$

Using Eq. (4.28), the quantity g/dp in Eq. (4.29) can be written as

$$\frac{g}{dp} = \frac{g}{-g\rho dz} = -\frac{1}{\rho dz} = -\frac{\rho_w/\rho}{\rho_w dz} = -\frac{q}{du} \tag{4.32}$$

where

$$q \equiv \frac{\rho_w}{\rho} = \frac{\text{density of water vapor}}{\text{density of air}}$$

is called the *specific humidity* of the air, and

$$du = \rho_w dz \tag{4.33}$$

is called the *density-weighted path length of water vapor* in the layer dz. Therefore, we can write Eq. (4.29) as

$$\frac{dT}{dt} = -\frac{q}{c_p} \frac{dE_\lambda(u)}{du} \tag{4.34}$$

If the solar spectrum is divided into n wavelength intervals, the total heating rate of a layer due to solar radiation can be written

$$\left(\frac{dT}{dt}\right)_{tot} = \sum_{i=1}^{n} \left(\frac{dT}{dt}\right)_i \tag{4.35}$$

Equations (4.34) and (4.35) can be used to compute the heating rate at various levels in the atmosphere due to the absorption of solar radiation by trace gases.

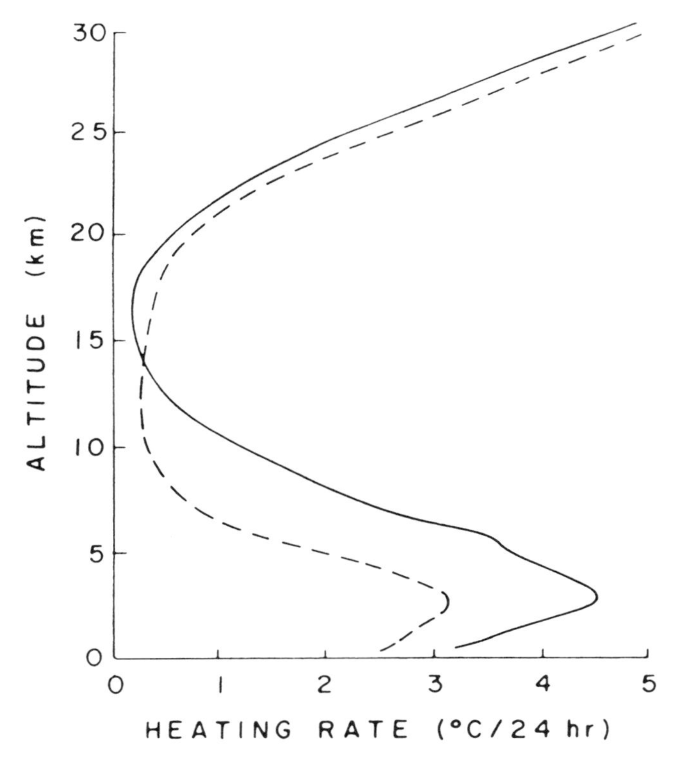

Figure 4.8. Calculated heating rates due to the absorption and multiple scattering of solar radiation by atmospheric gases in the tropics (solid line) and at middle latitudes (dashed line). The Sun is assumed to be overhead, and the surface of the Earth is assumed to reflect 15% of the incident solar radiation. [Adapted from *An Introduction to Atmospheric Radiation* by K.-N. Liou, Academic Press (1980).]

Figure 4.8 shows calculated solar heating rates of the atmosphere as a function of altitude for middle-latitudes and the tropics, assuming an overhead Sun for 24 hours. The calculations incorporate the effects of absorption by O_3, H_2O, O_2, and CO_2, scattering by gaseous molecules, and reflection from the Earth's surface. The higher concentrations of water vapor in the tropical troposphere cause greater heating rates in this region than in the troposphere at middle latitudes. Peaks in the heating rates, of about 3 to 4°C per 24 hours, occur at an altitude of about 3 km. The heating rate decreases sharply with height up to the tropopause (due primarily to the decrease in the amount of water vapor with height); the heating rate then increases again in the stratosphere. Above the tropopause, the increase in the rate of solar heating is due to absorption by stratospheric O_3.

4.5 Attenuation of solar radiation by aerosols

Aerosols, which derive from a variety of natural and anthropogenic sources (see Chapter 6), also scatter, and in some cases absorb, solar radiation. For example, it is estimated that the global mean optical depth due to aerosols at midvisible wavelength is ~0.1, with natural and anthropogenic aerosols contributing about equally. Unlike most gases in the atmosphere, aerosols are distributed very unevenly. In heavily polluted air the aerosol optical depth can be ~0.2 to 0.8, and in thick smoke (e.g., from biomass burning) it can be in excess of 1. The main natural contributors to the attenuation of solar radiation by aerosols are soil dust, sulfates, and organics; the main anthropogenic contributors are sulfates (from SO_2), possibly organics, and particles from biomass burning.

Because the most important aerosol scatterers have dimensions comparable to the wavelength of visible light, Rayleigh scattering does not apply in this case. Instead, the more complete theory of scattering for spherical particles developed by Mie[11] must be used. Shown in Figure 4.9 is the *scattering efficiency* K (namely, the ratio of the effective scattering cross section of a molecule or particle to its geometric cross section) versus the dimensionless quantity $\alpha \equiv 2\pi r/\lambda$, where r is the radius of the particle (assumed spherical) and λ the wavelength of the em radiation. As we have seen in Section 4.2, when $\alpha \ll 1$ Rayleigh scattering applies and K is proportional to α^4, which is represented by the first rapidly rising portion of the curve on the left-hand side of Figure 4.9. For the scattering of light by most aerosols, the values of α are between about 0.1 and 50 (the *Mie regime*). In this regime the value of K oscillates with α (Fig. 4.9), and forward scattering predominates over back scattering. If the particles are fairly uniform in size, the scattered sunlight is bluish or reddish in hue, depending on whether $dK/d\alpha$ is positive or negative at visible wavelengths. If the particles have a size spectrum broad enough to span several of the maxima and minima in the curve shown in Figure 4.9, the scattered light will be whitish.

When $\alpha > 50$, $K \simeq 2$ (the right-hand side of Fig. 4.9), and the angular distribution of the scattered radiation can then be described reasonably well by the theory of geometric optics. The scattering of visible radiation

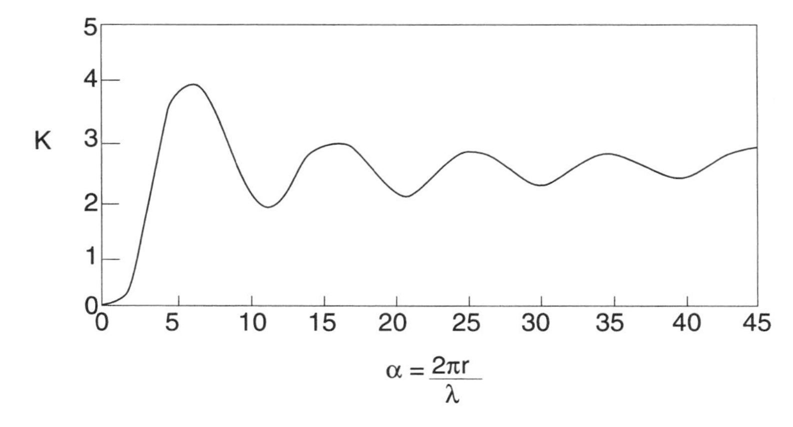

Figure 4.9. The scattering efficiency K versus the size parameter α for nonabsorbing spheres with a refractive index of 1.33.

by cloud droplets, raindrops, and ice particles falls within this regime and produces such optical phenomena as rainbows and haloes.

The main absorbers of solar radiation are carbonaceous particles (i.e., particles containing carbon), which are of two types: black carbon (also called *elemental* or *graphitic* carbon) and organic carbon. Black carbon, which is produced by combustion, is the most abundant light-absorbing aerosol in the atmosphere. Organic carbon aerosols are complex mixtures of many compounds that derive from both natural and anthropogenic sources.

4.6 Absorption and emission of longwave radiation

About 60% of the incoming solar radiation to the Earth is transmitted through the atmosphere, and about 40% is absorbed at the Earth's surface (Fig. 4.1). Since the Earth is approximately in thermal equilibrium it must, on average, lose the same amount of energy as it receives from the Sun. It does so by radiating energy upwards (as well as by the transfer of heat through evaporation and turbulent transfer) – see Figure 4.1. Most of the upward radiation is absorbed by the atmosphere which, in turn, emits radiation both upwards and downwards. Because of their relatively low temperatures (compared to the Sun), the surface of the Earth and the atmosphere emit most of their radiant energy at IR wavelengths, while most of the radiant energy from the Sun is in the visible and near IR portions of the em spectrum (see Fig. 4.2).

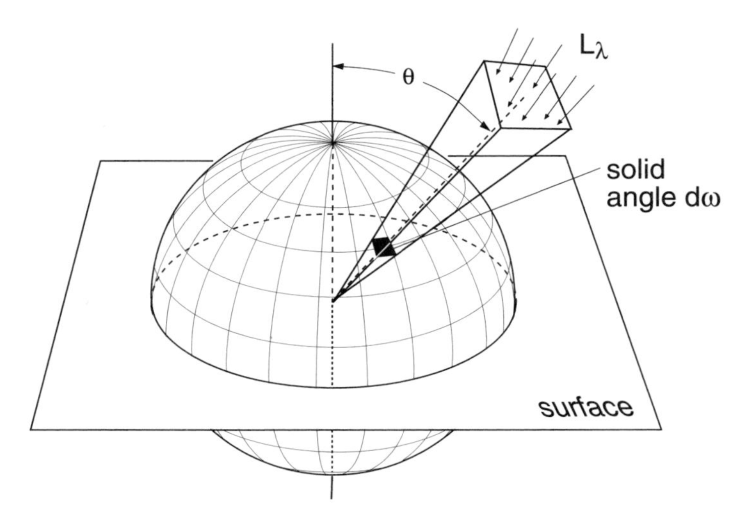

Figure 4.10. Relationship between radiance L_λ and irradiance E_λ. The irradiance of the surface is obtained by integrating the normal component of L^* to the surface over a hemisphere (see text).

There is another important difference between terrestrial and solar radiation. The solar radiation reaching the TOA can generally be considered as a parallel beam, whereas, terrestrial radiation within the Earth-atmosphere system can come from many different directions (so-called *diffuse radiation*). Consequently, calculations of terrestrial irradiances generally involve integrating IR radiation coming from all directions. To emphasize the processes that determine the emission and absorption of terrestrial radiation, we will set aside the geometrical problems associated with treating diffuse radiation. We can do this by using the various relationships we have derived for parallel beam solar radiation, provided we replace the irradiance E_λ by *radiance* L_λ. The radiance L_λ is defined as the irradiance (at wavelength λ) *per unit solid angle* (units are $\mathrm{W\,m^{-2}}$ per steradian per μm) reaching a surface from a specified infinitesimal arc of solid angle $d\omega$ (Fig. 4.10). If the direction of L_λ is at angle θ to the normal to the surface

$$E_\lambda = \int_0^{2\pi} L_\lambda \cos\theta \, d\omega \tag{4.36}$$

where the integration is carried out over a hemisphere (which subtends a solid angle 2π steradian). Another important difference in treating solar and terrestrial radiation is that in the latter case we must consider both the absorption and emission of longwave radiation, whereas emissions of shortwave radiation by the Earth and its atmosphere are negligible (Fig. 4.2).

The absorption of terrestrial radiation along an *upward* path through the atmosphere is described by a similar expression to Eq. (4.6) but with the sign reversed and with E_λ replaced by L_λ

$$-dL_\lambda = L_\lambda \, \alpha_{a\lambda} \, \rho \, dz \tag{4.37}$$

where we have used Eq. (4.8) to substitute for $b_{a\lambda}$ and dz is the path length along the line of the radiation. The emission of radiation from a gas can be treated in a similar manner to the way we treated absorption in the previous section. Thus, an analogous expression to Eq. (4.37) for emission is

$$dL_\lambda = L_\lambda^* d\varepsilon_\lambda = L_\lambda^* da_\lambda = L_\lambda^* \alpha_{a\lambda} \rho dz \tag{4.38}$$

where ε_λ is the emissivity of the gas, which is equal to its absorptivity a_λ (Kirchhoff's law), and L_λ^* is the blackbody radiance and temperature $T(z)$.

The net contribution of the layer of thickness dz to the radiance of wavelength λ passing upward through it is the difference between the emission (given by Eq. (4.38)) and the absorption (given by Eq. (4.37)). Therefore,

$$dL_\lambda(\text{net}) = L_\lambda^* \alpha_{a\lambda} \rho \sec\theta \, dz - L_\lambda \alpha_{a\lambda} \rho \sec\theta \, dz$$

or

$$dL_\lambda(\text{net}) = -\alpha_{a\lambda}\left(L_\lambda - L_\lambda^*\right) \rho \sec\theta \, dz \tag{4.39}$$

where dz is now the path length normal to the surface (see Fig. 4.10). Equation (4.39), which is known as *Schwarzschild's*[12] *equation*, is the basis for computing the transfer of longwave radiation. For an isothermal atmosphere, L_λ^* is constant. In this case, integration of Eq. (4.39) yields

$$\int_{L_{\lambda 0}}^{L_\lambda} \frac{dL_\lambda}{L_\lambda - L_\lambda^*} = -\int_0^z \alpha_{a\lambda} \rho \sec\theta \, dz$$

or, using Eqs. (4.8) and (4.10),

$$\ln\left(\frac{L_\lambda - L_\lambda^*}{L_{\lambda 0} - L_\lambda^*}\right) = -\tau_\lambda \sec\theta$$

where τ_λ is the optical depth from the Earth's surface ($z = 0$) to height z, and $L_{\lambda 0}$ and L_λ and the upward radiances at $z = 0$ and at z. Rearrangement of the last expression yields

$$L_\lambda - L_\lambda^* = (L_{\lambda 0} - L_\lambda^*)\exp(-\tau_\lambda \sec\theta) \tag{4.40}$$

Equation (4.40) shows that L_λ approaches L_λ^* as the optical depth τ_λ increases. That is, a very deep atmosphere radiates like a blackbody.

4.7 The greenhouse effect, radiative forcing, and global warming

As we have seen, about 50% of the solar radiation incident on the TOA reaches the surface of the Earth, where it is absorbed. Consequently, the Earth's surface is the principal source of heat for the atmosphere. The longwave radiation from the Earth's surface is largely absorbed by the atmosphere. The atmosphere, in turn, emits longwave radiation upward and downward in amounts that vary with the temperature of the air at the level of the emission. The downward flux of longwave radiation from the atmosphere causes the surface of the Earth to have a higher temperature than it would have in the absence of an absorbing atmosphere. This is called the *greenhouse effect*. The following exercise illustrates how this effect comes about for a simplified Earth–atmosphere system.

Exercise 4.4. (a) The *emission temperature* T_E of the Earth is defined as the temperature it would have to have to achieve radiative energy balance (i.e., solar radiation absorbed = terrestrial radiation emitted) assuming that the Earth radiates as a blackbody. Derive an expression for the emission temperature of the Earth, assuming that it does *not* have an atmosphere, in terms of the solar irradiance at the TOA (E_∞), the radius of the Earth (R_E), and the fraction of the incoming solar radiation reflected back to space by the Earth (R, called the *albedo* of the Earth). If $E_\infty = 1{,}368\,\mathrm{W\,m^{-2}}$, $R = 0.3$, and $\sigma = 5.67 \times 10^{-8}\,\mathrm{W\,m^{-2}\,deg^{-4}}$, what is the value of T_E for this hypothetical Earth?

(b) Now consider a more realistic approximation to the Earth–atmosphere system, namely, an atmosphere that is blackbody (i.e., opaque) for terrestrial radiation but transparent for solar radiation. Show that in this case the surface temperature of the Earth (T_s) is greater than T_E by about 20%.

Solution. (a) The radiant flux from the Sun incident on a surface oriented normal to the Sun's parallel rays and located at the same distance from the Sun as the Earth is E_∞ W m^{-2}. When the hemisphere of the Earth that is facing the Sun at any given time is projected onto a plane normal to the Sun's rays it has an area πR_E^2. Therefore, the total radiant energy incident on the Earth from the Sun is $E_\infty \pi R_E^2$ watts. However, a fraction R of this energy is reflected back to space. Therefore, the radiant energy from the Sun that is absorbed by the Earth is $E_\infty (1 - R) \pi R_E^2$ watts.

For the radiative energy balance of the Earth

solar radiation absorbed = terrestrial radiation emitted

Therefore, from the definition of T_E,

$$E_\infty (1 - R) \pi R_E^2 = \sigma T_E^4 4\pi R_E^2$$

where we have used the Stefan–Boltzmann law (Eq. 4.1), and the fact that the total surface area of the Earth is $4\pi R_E^2$. Therefore,

$$\frac{E_\infty}{4}(1 - R) = \sigma T_E^4 \tag{4.41}$$

Substituting E_∞ = 1,368 W m^{-2}, R = 0.3 and $\sigma = 5.67 \times 10^{-8}$ W m^{-2} deg^{-4} into this expression yields $T_E = 255$ K, or –18°C. Even though the emission temperature is not the actual temperature of this hypothetical Earth, the value of T_E we have calculated is far below the actual global mean surface temperature of the real Earth (15°C).

(b) In this case, the energy balance at the TOA is still given by Eq. (4.41). However, since the atmosphere absorbs all of the longwave radiation from the Earth's surface, the only radiation emitted to space by the Earth–atmosphere system is from the atmosphere. Therefore, the energy balance at the TOA is now (Fig. 4.11)

$$\frac{E_\infty}{4}(1 - R) = \sigma T_A^4 = \sigma T_E^4 \tag{4.42}$$

We see from Eq. (4.42) that $T_A = T_E$. However, the temperature of the surface of the Earth (T_s) is greater than T_A (or T_E) as we can see if we write down the energy balance for the atmosphere itself (Fig. 4.11)

$$\sigma T_s^4 = 2\sigma\, T_A^4$$

or

$$T_S = 1.19\, T_A \tag{4.43}$$

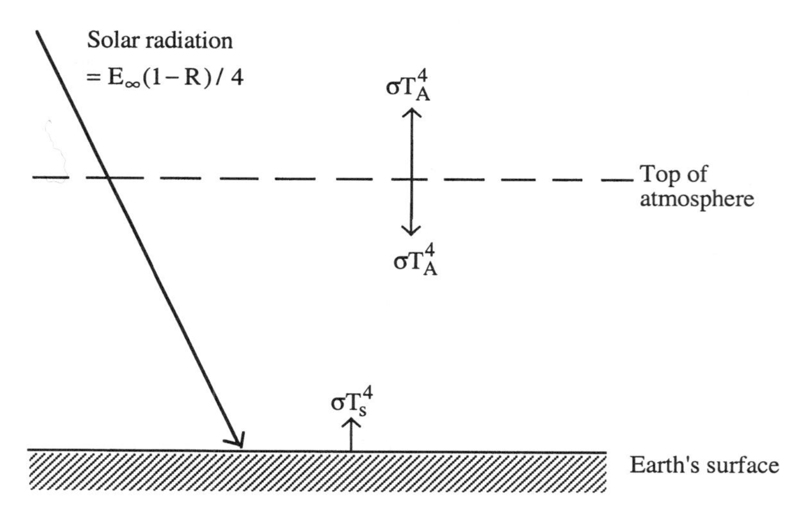

Figure 4.11. Irradiances for the Earth–atmosphere system assuming that the atmosphere is transparent to solar radiation but opaque to terrestrial radiation.

Changes in the concentrations of greenhouse gases (and aerosols) in the atmosphere have led to a perturbation in the radiation balance of the Earth–atmosphere system, which is referred to as a *radiative forcing.*[13] Radiative forcing can be defined as the perturbation to the net irradiance (in $W m^{-2}$) at the TOA.[14] If the perturbation is positive (i.e., there is an increase in the net irradiance to the Earth–atmosphere system at the TOA), it will tend to increase the temperature of the Earth–atmosphere system; if the perturbation is negative, it will tend to lower the temperature of the Earth–atmosphere system.

The globally averaged radiative forcing due to changes in the concentrations of the well-mixed greenhouse gases (CO_2, CH_4, N_2O, and halocarbons) since preindustrial times is estimated to be $+2.45 \pm 0.37 W m^{-2}$. Increases in the concentrations of CO_2, CH_4, and N_2O of about 30%, 145%, and 15%, respectively, account for contributions of 64%, 19%, and 6%, respectively, to this radiative forcing. An increase in tropospheric O_3 (not well mixed) is estimated to have caused an additional +0.2 to $+0.6 W m^{-2}$ of radiative forcing. Since many greenhouse gases have long residence times in the atmosphere (e.g., CO_2 and N_2O), they produce long-term radiative forcing.[15] The contribution to the direct radiative forcing by halocarbons (i.e., chlorofluorocarbons, CFC, and hydrochlorofluorocarbons, HCFC) is about $+0.25 W m^{-2}$. However, the net radiative forcing due to halocarbons is reduced to about 0.1 W m^{-2} because they have caused reductions in the concentration of ozone in the stratosphere[16] (see Section 10.2).

In contrast to greenhouse gases, increases in atmospheric aerosol concentrations tend to cool the atmosphere by increasing the reflection of solar radiation back to space. However, also in contrast to the long-lived and well-mixed greenhouse gases, atmospheric aerosols from anthropogenic sources have short residence times, and they tend to be concentrated in industrialized/urban areas, where they produce large but local radiative forcing. For example, it has been estimated that the average radiative forcing in the Amazon Basin during the peak two months of the biomass burning season is -15 ± 5 W m^{-2}. However, the globally-averaged radiative forcing due to smoke from biomass burning is only about -0.3 W m^{-2}. The total globally averaged direct radiative forcing due to anthropogenic aerosols is estimated to be about -0.5 W m^{-2}. Aerosols may also have an important indirect radiative effect due to the fact that a component of the aerosol (called *cloud condensation nuclei*) can increase the concentration of cloud droplets and thereby increase the reflection of solar radiation by clouds. However, the negative radiative forcing due to this indirect effect of aerosols is poorly quantified.

Figure 4.12 summarizes and compares estimates of the globally and annually averaged radiative forcings due to changes in the concentrations of greenhouse gases and aerosols from preindustrial times to 1995.

4.8 Photochemical reactions[17]

Photochemical reactions (i.e., reactions driven by the interaction of a photon of electromagnetic radiation and a molecule), which are referred to as *photolysis* if the molecule dissociates, play a key role in many aspects of atmospheric chemistry. Table 4.1 lists some examples of these reactions.

Whether a molecule can be involved in a photochemical reaction depends on the probability of it absorbing a photon with sufficient energy to cause dissociation of the molecule. This will depend, in part, on the radiant flux *from all directions* incident on a volume of the air, which is called the *actinic flux* (I). The common (practical) units for I are photons cm^{-2} s^{-1} or, if we wish to specify a particular wavelength interval, photons cm^{-2} s^{-1} μm^{-1}. Comparing these definitions and units with

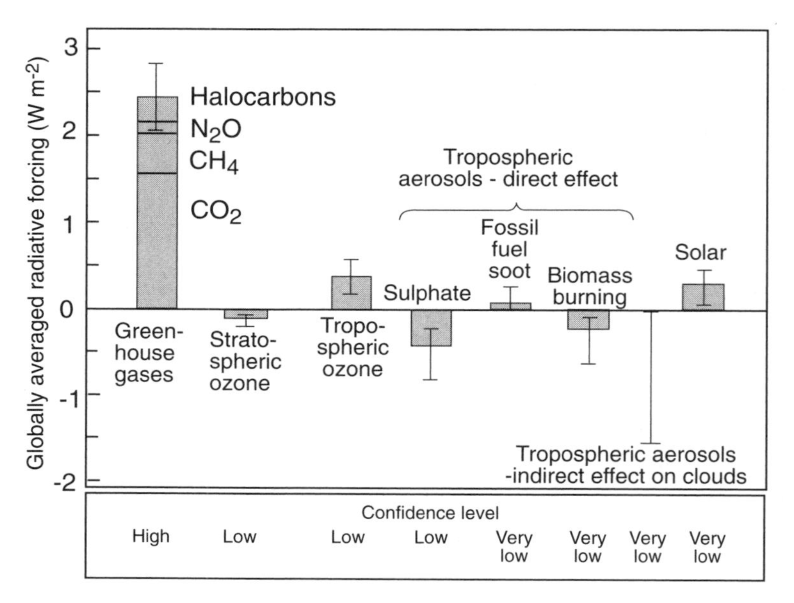

Figure 4.12. Estimate of the globally and annually averaged anthropogenic radiative forcings due to changes in the concentrations of greenhouse gases and aerosols from preindustrial times to 1995 and to natural changes in solar output from 1850. The height of the rectangular bar indicates a midrange estimate of the forcing, and the error bar shows an estimate of the uncertainty range. A subjective confidence that the actual forcing lies within the error bar is indicated by the "confidence level." [From *Climate Change 1995*, Ed. J. T. Houghton et al., Cambridge University Press (1996).]

those given in Section 4.1, we see that actinic flux is similar (but not identical) to irradiance.

For a photochemical reaction involving a species A, the rate of dissociation of A can be written

$$\frac{d[\mathrm{A}]}{dt} = -j_{\mathrm{A}}[\mathrm{A}] \tag{4.44}$$

where [A] is the number concentration of the molecules of A and j_{A} is the *first-order photolysis rate coefficient* (units are s^{-1}) for the reaction. The magnitude of j_{A} depends on the probability that a photon will be absorbed by a molecule of A (as measured by the *absorption cross section* σ_{A} (cm^{-2}) of a molecule of A), the probability that if absorption occurs the molecule A will dissociate (as measured by the *quantum yield* ϕ_{A}), and on the actinic flux I. Therefore, if we consider the wavelength

Table 4.1. *A few examples of the many photochemical reactions of importance in atmospheric chemistry*

Reaction[a]	Comments
$6H_2O + 6CO_2 + h\nu \rightarrow 6O_2 + C_6H_{12}O_6$	Photosynthesis by green plants (see Section 1.3).
$NO_2 + h\nu \rightarrow NO + O$	Occurs for wavelengths from 0.400 to 0.625 μm. The oxygen atom produced by this reaction leads to the only *in situ* chemical source of O_3 in the troposphere: $O_2 + O + M \rightarrow O_3 + M$ (see Sections 5.2 and 9.2).
$HONO + h\nu \rightarrow OH + NO$	A source of the important hydroxyl (OH) radical. Occurs at relatively long wavelengths (<0.400 μm) that reach ground level.
$HCHO + h\nu \rightarrow H + HCO$	The photolysis of formaldehyde (HCHO) is a significant source of free radicals in the troposphere (see Section 5.2). Occurs at wavelengths <0.340 μm.
$O_2 + h\nu \rightarrow O + O$	The first of the Chapman reactions (see Section 10.1). Occurs in the stratosphere due to the absorption of solar radiation in the 0.2 to 0.22 μm and 0.185 to 0.2 μm wavelength regions.
$O_3 + h\nu \rightarrow O_2 + O$	The third of the Chapman reactions (see Section 10.1). Occurs in the stratosphere (and troposphere) due to the absorption of solar radiation in the ~0.305 to 0.320 μm wavelength range (see Section 5.2).
$CFCl_3 + h\nu \rightarrow CFCl_2 + Cl$	Example of photolysis of chlorofluorocarbons in the stratosphere. Occurs at wavelengths from 0.19 to 0.22 μm (see Section 10.2).

[a] $h\nu$ represents 1 photon.

Table 4.2. *Values of the quantum yields and absorption cross sections for Reaction (4.46) at 25°C for various wavelength bands between 0.295 μm and 0.410 μm; also listed is the actinic flux at the surface at solar noon at a latitude of 40°N on 1 March*

Wavelength Interval (μm)	Quantum Yield	Absorption Cross Section (in units of 10^{-19} cm^2)	Actinic Flux (photons cm^{-2} s^{-1})
0.295–0.300	0.98	1.07	2.00×10^{11}
0.300–0.305	0.98	1.42	7.80×10^{12}
0.305–0.310	0.97	1.71	4.76×10^{13}
0.310–0.315	0.96	2.01	1.44×10^{14}
0.315–0.320	0.95	2.40	2.33×10^{14}
0.320–0.325	0.94	2.67	3.18×10^{14}
0.325–0.330	0.93	2.89	4.83×10^{14}
0.330–0.335	0.92	3.22	5.30×10^{14}
0.335–0.340	0.91	3.67	5.37×10^{14}
0.340–0.345	0.90	3.98	5.90×10^{14}
0.345–0.350	0.89	4.09	5.98×10^{14}
0.350–0.355	0.88	4.62	6.86×10^{14}
0.355–0.360	0.87	4.82	6.39×10^{14}
0.360–0.365	0.86	5.15	7.16×10^{14}
0.365–0.370	0.85	5.60	8.90×10^{14}
0.370–0.375	0.84	5.39	8.11×10^{14}
0.375–0.380	0.83	5.67	9.14×10^{14}
0.380–0.385	0.82	5.97	7.63×10^{14}
0.385–0.390	0.80	5.97	8.46×10^{14}
0.390–0.395	0.77	5.95	8.78×10^{14}
0.395–0.400	0.75	6.33	1.07×10^{15}
0.400–0.405	0.55	6.54	1.28×10^{15}
0.405–0.410	0.23	6.05	1.44×10^{15}

and temperature dependence of these various parameters, the value of j_A for wavelengths between λ_1 and λ_2 is

$$j_A = \int_{\lambda_1}^{\lambda_2} \sigma_A(\lambda, T) \phi_A(\lambda, T) I(\lambda) d\lambda \tag{4.45}$$

where the units of I are now photons $cm^{-2}s^{-1}\mu m^{-1}$, if $d\lambda$ is expressed in μm. Note that σ_A and ϕ_A are fundamental properties of the molecule A, which can be determined from laboratory experiments.

Exercise 4.5. Using the information given in Table 4.2, calculate the value of the photolysis rate coefficient j between wavelengths of 0.295 and 0.410 μm for the reaction

$$NO_2 + h\nu \rightarrow NO + O \tag{4.46}$$

Assume a cloud-free day and a surface albedo of zero.

Solution. The right-hand side of Eq. (4.45) can be approximated by summing over small wavelength intervals. Hence, for this exercise,

$$j = \sum_i \bar{\sigma}_A(\lambda_i, T)\bar{\phi}_A(\lambda_i, T)\bar{I}(\lambda_i) \tag{4.47}$$

where the overbars indicate average values over the wavelength interval $\Delta\lambda_i$ centered on wavelength λ_i.

Using the values given in Table 4.2, and carrying out the summation of products on the right-hand side of Eq. (4.47), yields a value of j of $5.50 \times 10^{-3}\,s^{-1}$ for Reaction (4.46) under the specified conditions.

Exercises

See Exercises 1(j)–(n) and Exercises 14–17 in appendix I.

Notes

1 For a more detailed discussion of the basic principles of atmospheric radiation, the reader is referred to Chapters 6 and 7 of *Atmospheric Science: An Introductory Survey* by J. M. Wallace and P. V. Hobbs (Academic Press, New York, 1977).

2 Josef Stefan (1835–1893). Austrian physicist. Became professor of physics at the University of Vienna at age 28. Originated the theory of diffusion of gases and carried out fundamental studies on radiation.

3 Wilheim Wien (1864–1928). German physicist. Received the Nobel Prize in 1911 for the discovery (in 1893) of the displacement law named after him. Also made the first (rough) determination of the wavelength of x-rays.

4 Gustav Kirchhoff (1824–1887). German physicist. In addition to his work in radiation, he made fundamental discoveries in electricity and spectroscopy. Developed (with Bunsen) spectrum analysis. Discovered cesium and rubidium.

5 August Beer (1825–1863). German physicist, noted for his studies of optics.

6 Lord Rayleigh (John William Strutt, 3rd Baron) (1842–1919). English mathematician and physicist. Best known for his work on the theory of sound and the scattering of light. Discovered (with Ramsey) the presence of argon in the air, for which he won the Nobel Prize in 1904.

7 The refractive index is difficult to define for a molecule. The approach is to consider an ensemble of molecules, for which the scattering may be related to the macroscopic index of refraction of the gas.

8 Some of the more important absorption bands in the solar spectrum are given names. For example, the absorption bands due to O_3 in the near-UV at wavelengths between 0.3 and 0.36 μm (which can be seen on the left-hand side of Fig. 4.6) and a series of bands centered at 0.255 μm, are called the *Huggins* and *Hartley* bands, respectively. Although the Hartley bands are quite weak, they appear in the solar spectrum when the Sun is low in the sky; they were responsible for the first positive identification of O_3 in the atmosphere. The absorption band due to O_3 at visible wavelengths is called the *Chappuis* band, and the O_2 band at 0.7 μm is called the *A-band*.

9 The less mathematically inclined reader may wish to skip Sections 4.3, 4.4, and the quantitative portions of Section 4.6.

10 For a derivation and discussion of the hydrostatic equation and the dry adiabatic lapse rate see Chapter 2 of *Atmospheric Science: An Introductory Survey* by J. M. Wallace and P. V. Hobbs (Academic Press, New York, 1977).

11 Gustav Mie (1868–1957). German physicist. Carried out fundamental studies on the theory of em scattering and kinetic theory.

12 Karl Schwarzschild (1873–1916). German astronomer skilled in both theoretical and experimental work, and also in popularizing astronomy. In 1960 the Berlin Academy referred to him as "the greatest German astronomer of the last hundred years."

13 Radiative forcing provides a quantitative estimate of the potential impact of changes in the composition of the Earth's atmosphere on the climate of the Earth. However, because the concept of "radiative forcing" does not account for feedbacks, general circulation models (GCM) of the atmosphere are needed to better estimate the many possible effects on climate of various types of radiative forcing.

14 A stricter definition of radiative forcing is the perturbation in the net irradiance (in $W m^{-2}$) at the tropopause after allowing for stratospheric temperatures to adjust to radiative equilibrium (which takes a few months), but with surface and tropospheric temperatures and atmospheric moisture held constant.

15 If CO_2 emissions remain at their levels at the end of the twentieth century, atmospheric CO_2 concentrations will reach about 500 ppmv (i.e., about twice their preindustrial concentrations) by the end of the twenty-first century.

16 Radiative forcing due to changes in stratospheric O_3 is difficult to calculate because such changes cause a significant change in both solar and terrestrial radiation. Also, the depletion of O_3 causes changes in stratospheric temperatures, which significantly modify radiative forcing. Finally, radiative forcing is sensitive to the spatial (particularly the vertical) distribution of perturbations in O_3.

17 Some of the basic principles of photochemistry are discussed in Chapter 7 of *Basic Physical Chemistry for the Atmospheric Sciences* by P. V. Hobbs (Cambridge University Press, New York, 1995).

5

Sources, transformations, transport, and sinks of chemicals in the troposphere

In this chapter we will consider the sources, transformations, transport, and sinks of chemicals in the natural troposphere. Our emphasis will be on gases; aerosols will be considered in more detail in Chapter 6.

5.1 Sources

The principal natural sources of gases in the troposphere are the biosphere, the solid Earth, the oceans, and *in situ* formation in air from other chemical species. These sources are discussed, in turn, in the following subsections.

a. Biological

Even though the biosphere contains only a small fraction of the total reservoirs of most chemicals on Earth, it plays a major role in determining the abundances and transport of many gases in the atmosphere. The smells associated with flowers and other forms of vegetation provide direct evidence that biota emit chemicals into the air. Some important biological sources of trace gases that enter the atmosphere are:

- Photosynthesis in plants (Reaction (1.1)), which is responsible for virtually all of the oxygen in the atmosphere.
- Respiration (the reverse of Reaction (1.1)), which releases CO_2 into the air. This can be seen, on a seasonal basis, in Figure 1.1 where the decline of CO_2 in the summer months is due to its uptake by plants during photosynthesis. The rise of CO_2 in winter and early spring (Fig. 1.1) is due to respiration, and the decay of leaf litter and other dead plant material. This "breathing" of the atmosphere can also be

detected in diurnal fluctuations in CO_2; CO_2 concentrations in forests can be ~35 ppmv higher at night than in the day.

- Methane (CH_4), the major hydrocarbon in air, about 80% of which derives from recent organic materials (rather than fossil fuels). Cud-chewing animals (cows etc.), termites, rice paddies, tundra, and wetlands (marshes) are the major sources of CH_4.
- Terpenes (a class of hydrocarbons that derives from the isoprene unit, C_5H_8) evaporate from leaves. About 80% of these terpenes oxidize to organic aerosols in an hour or so. Emissions from vegetation are a significant source of hydrocarbons, which can react photochemically, together with NO_x (i.e., NO and NO_2) gases, to produce O_3. Ozone plays a central role in atmospheric chemistry (see Sections 5.2 and 10.1).
- Other gases that derive in part from the biosphere (e.g., CO and N_2O) are also involved in the control of O_3.
- Biomass burning releases many gases into the air (e.g., CO_2, CO, CH_4, H_2, N_2, NMHC, NO, NO_2, COS, CH_3Cl). Many photochemically reactive species are generated by biomass burning, which give rise to O_3 production and photochemical smog-like species (e.g., peroxyacetyl nitrate ($CH_3COO_2NO_2$), or PAN for short). Much biomass burning is anthropogenic in origin.
- The biological transformation (often by microbes) of N_2 into NH_3 (primarily from animal urine and soils), N_2O (through nitrate respiration, performed by aerobic bacteria in soils), and NO.
- Oceanic regions of high biological productivity and organic content, particularly upwelling regions, coastal waters and coastal salt marshes, are the major source of carbon disulphide (CS_2) and carbonyl sulfide (COS); the latter is the most abundant species of gaseous sulfur in the troposphere. Phytoplankton are the major sources of dimethylsulfide (CH_3SCH_3, or DMS for short), and dimethyldisulfide (CH_3SSCH_3) in the atmosphere.[1] DMS is oxidized to SO_2 and then to sulfate aerosols. Microbial degradation of dead organic matter liberates hydrogen sulfide (H_2S). Thus, saline marshlands and anaerobic (i.e., oxygen-poor) swamps are appreciable sources of H_2S, which, in part, gives these areas their characteristic rotten egg odor.
- Important sources of methyl chloride (CH_3Cl), which is the most abundant halocarbon in air, include biological activity in seawater, wood molds, and biomass burning. Methyl chloride is the major natural source of stratospheric chlorine.
- Molecular hydrogen is emitted by microbiological activity in the

oceans, biomass burning, the fermentation of bacteria in soils, and photolysis of formaldehyde.

- Uses by humans of biogenic materials result in the emissions of numerous chemicals into the atmosphere – CO_2, CO, NO_x, N_2O, NH_3, SO_2, and HCl (from the combustion of oil, gas, coal, and wood), hydrocarbons (from motor vehicles, refineries, paints, solvents, etc.), H_2S (from paper mills, oil refineries, and animal manure), COS (from natural gas), methyl mercaptan (CH_3SH) and DMS (from animal manure, paper mills, and oil refining), HCl (from coal combustion), CH_3Cl (from tobacco burning), chloroform, $CHCl_3$ (from combustion of petroleum and bleaching of woods), and many others.

Exercise 5.1. A simple model for the exchange of COS between the surface layer of the ocean, where it is generated, and the lower regions of the ocean could be based on the following assumptions. (i) COS in the surface layers of the ocean is in equilibrium with COS in the atmosphere and there is no net atmospheric sink. (ii) The concentration of COS in the surface layer of the ocean, C(0), is constant at $1.0 \times 10^{-11}\,\mathrm{kg\,m^{-3}}$. (iii) Beneath the surface layer of the ocean COS is destroyed chemically with a first-order rate coefficient $k = 5.0 \times 10^{-6}\,\mathrm{s^{-1}}$. (iv) Beneath the same surface layer, the chemical destruction of COS is balanced by the downward transport of eddies with a diffusion coefficient of $0.0050\,\mathrm{m^2\,s^{-1}}$. (v) The ocean is infinitely deep and uniform in the horizontal. Use these assumptions to determine:

(a) At what depth the concentration of COS in the ocean is equal to C(0)/2.
(b) The column density of COS in the ocean.
(c) The average lifetime of a COS molecule in the ocean.

Solution. Since the ocean is uniform in the horizontal, we need only consider transfer in the vertical (z) direction (i.e., we can use a one-dimensional (1D) model). Consider a small slab of the ocean at distance z below the ocean surface that has a unit cross-sectional area and thickness dz (Fig. 5.1). Let $F(z)$ be the flux of COS into the top of this slab and $F(z + dz)$ the flux out of the base of the slab. The net flux of COS into the slab is

$$\Delta F = F(z) - F(z + dz) = F(z) - \left[F(z) + \frac{dF(z)}{dz} dz \right] = -dF(z)$$

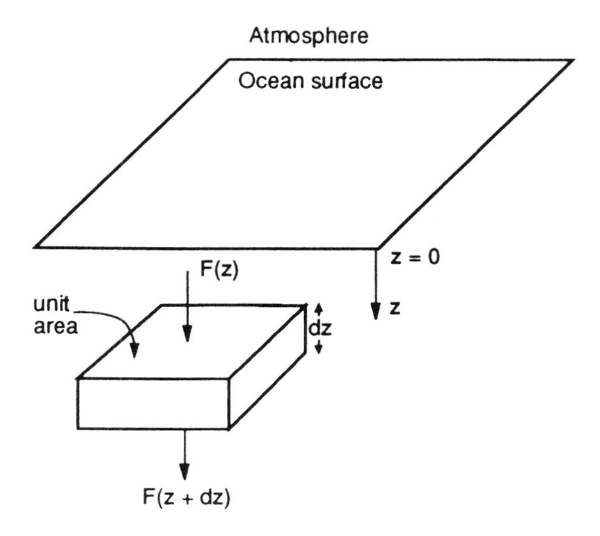

Figure 5.1. Sketch for Exercise 5.1.

Since the volume of the slab is $1 \cdot dz$

$$\text{net flux of COS into a unit volume of the slab} = -\frac{dF(z)}{dz} \quad (5.1)$$

If D is the eddy diffusion coefficient,

$$F(z) = -D\frac{dC(z)}{dz} \quad (5.2)$$

where $C(z)$ is the concentration of COS at distance z below the ocean surface, and the negative sign arises because diffusion occurs in the direction opposite to that of increasing concentration. In the present case, the concentration of COS is greatest at the ocean surface and decreases downward; therefore, the flux of COS is downward in the ocean. From Eqs. (5.1) and (5.2)

$$\text{net flux of COS into a unit volume of the slab} = D\frac{d^2C(z)}{dz^2} \quad (5.3)$$

If k is the first-order rate coefficient for the chemical destruction of COS in seawater

$$\text{rate of destruction of COS in a unit volume of the slab} = kC(z) \quad (5.4)$$

Under steady-state conditions, from Eqs. (5.3) and (5.4)

$$\frac{dC(z)}{dt} = D\frac{d^2C(z)}{dz^2} - kC(z) = 0 \tag{5.5}$$

The solution to Eq. (5.5) is

$$C(z) = C(0)\exp(-z/H) \tag{5.6}$$

where H (a scale height) is given by

$$H = \left(\frac{D}{k}\right)^{\frac{1}{2}} = \left(\frac{0.0050}{5.0 \times 10^{-6}}\right)^{\frac{1}{2}} = 32\,\text{m} \tag{5.7}$$

We can now answer the three questions.

(a) When $C(z) = C(0)/2$, we have from Eqs. (5.6) and (5.7)

$$\frac{1}{2} = \exp\left(-\frac{z}{H}\right) = \exp\left(-\frac{z}{32}\right)$$

Therefore, $z = 22\,\text{m}$.

(b) The column density of COS in the ocean is the total mass of COS in a column of unit cross-sectional area extending from the surface to the bottom of the ocean. Therefore,

$$\begin{aligned}
\text{column density of COS} &= \int_{z=0}^{\infty} C(z)dz \\
&= \int_0^{\infty} C(0)\exp\left(-\frac{z}{H}\right)dz \\
&= C(0)H \\
&= \left(1.0 \times 10^{-11}\right) \times 32 \\
&= 3.2 \times 10^{-10}\,\text{kg m}^{-2}
\end{aligned}$$

(c) From Eq. (2.4)

$$\text{lifetime of a chemical } (\tau) = \frac{M}{F}$$

where M is the amount of the chemical in, say, the column of unit cross-sectional area we considered earlier and F is the efflux (rate of removal plus rate of destruction) of the chemical from the column. Therefore, the average lifetime of a COS molecule in the ocean is given by

$$\begin{aligned}\tau &= \frac{C(0)H}{k\int_0^\infty C(z)dz} \\ &= \frac{C(0)H}{kC(0)H} \\ &= \frac{1}{k} \\ &= \frac{1}{5.0\times10^{-6}}\,\text{s} \\ &= 2.3 \text{ days}\end{aligned}$$

b. Solid Earth

Volcanoes are the most important geochemical source of trace constituents in the atmosphere. Although highly localized and very variable, when volcanic emissions are blasted into the stratosphere (or higher) they can be rapidly dispersed around the globe and have long residence times (1 to 2 a). The violent eruption of Krakatoa in 1863, which was one of the largest volcanic eruptions witnessed by humans, propelled ash to a height of ~80 km (see Fig. 3.1). The fine dust drifted several times around the Earth. Aerosols produced from the gases remained in the stratosphere for several years. In 1991 the eruption of Mount Pinatubo in the Philippines produced large, transient increases in atmospheric aerosols. This, in turn, resulted in a global mean cooling at the surface of ~0.4°C in the year following the eruption; this cooling disappeared after ~3 years as the dust in the atmosphere diminished.

In addition to ash and dust particles, volcanoes emit H_2O, CO_2, SO_2, H_2S, COS, CS_2, HCl, HF, HBr, CH_4, CH_3Cl, H_2, CO, and several volatile heavy metals (e.g., Hg). As far as the budgets of trace atmospheric constituents are concerned, volcanoes contribute the most to sulfur gases (~5%). Although this percentage is small, volcanic emissions make an important contribution to sulfur in the stratosphere (see Section 10.3).

Rocks are a source of small quantities of certain gases, and they are the major sources of the gases He, Ar, and Rn in the atmosphere. Helium is produced primarily from the radioactive decay of uranium-238 and thorium-232. It does not accumulate in the atmosphere because it is so light that it escapes from the exosphere (see Section 3.3). Argon has accumulated over eons from the radioactive decay of potassium-40 in rocks. (The abundance of atmospheric argon can be used to calculate an approximate age for the Earth of about 4.5 Ga.) Radon-222 is a decay product of uranium in rocks; it has a half-life of only 3.8 days.

Carbonate rocks, such as limestone, which occur mainly as calcite (i.e., $CaCO_3$), contain 100,000 times more carbon than the atmosphere (see Table 1.2), but most of this is sequestered. However, carbonate rocks and marine sediments are involved in a long-period cycle with atmosphere CO_2 as follows. The weathering of calcite by CO_2 dissolved in rainfall and fresh waters (rivers and lakes) can be represented by

$$CO_2(g) + H_2O(l) \rightleftarrows CO_2(aq) + H_2O(l) \quad (5.8a)$$

$$2H_2O(l) + CO_2(aq) \rightleftarrows H_3O^+(aq) + HCO_3^-(aq) \quad (5.8b)$$

$$CaCO_3(s) + H_3O^+(aq) + HCO_3^-(aq) \rightleftarrows Ca^{2+}(aq) + 2HCO_3^-(aq) + H_2O(l) \quad (5.8c)$$

$$\text{Net:} \quad CaCO_3(s) + CO_2(g) + H_2O(l) \rightleftarrows Ca^{2+}(aq) + 2HCO_3^-(aq) \quad (5.9)$$

Reaction (5.8a) represents the equilibrium between CO_2 in air and in fresh water rivers and lakes.[2] In Reaction (5.8b), CO_2 receives an OH^- from H_2O to form the very weak (i.e., little ionized) acid HCO_3^- (the bio-carbonate ion), and in Reaction (5.8c) the H_3O^+ takes CO_3^{2-} away from $CaCO_3$ to form another bicarbonate ion. The forward Reaction of (5.9) represents the weathering of calcite by CO_2 dissolved in fresh waters.

The weathering products on the right side of Reaction (5.9) eventually enter the oceans, where they precipitate to form new sediments (the reverse of Reaction (5.9)). Through uplift of continental shelf regions, subduction of marine sediments into the upper mantle and lower crust of the Earth, and volcanic eruptions, these products are eventually returned to continental sediments, thereby completing this geochemical cycle. The residence times of $Ca^{2+}(aq)$ and $HCO_3^-(aq)$ in the oceans are $\sim 8 \times 10^5$ and $\sim 7.5 \times 10^4$ years, respectively.

Exercise 5.2. When CO_2 dissolves in pure water the following reactions occur

$$CO_2(g) + H_2O(l) \rightleftarrows H_2CO_3(aq) \quad (5.10a)$$

$$H_2CO_3(aq) + H_2O(l) \rightleftarrows HCO_3^-(aq) + H_3O^+(aq) \quad (5.10b)$$

$$HCO_3^-(aq) + H_2O(l) \rightleftarrows CO_3^{2-}(aq) + H_3O^+(aq) \quad (5.10c)$$

$$\text{Net:} \quad CO_2(g) + 3H_2O(l) \rightleftarrows CO_3^{2-}(aq) + 2H_3O^+(aq) \quad (5.11)$$

where $H_2CO_3(aq)$ is carbonic acid, which, since it is diprotic (i.e., contributes two protons to water), equilibrates with H_2O to form a low concentration of the bicarbonate ion (Reaction (5.10b)). The bicarbonate ion equilibrates with H_2O to form a very low concentration of $CO_3^{2-}(aq)$ (Reaction (5.10c)). The acid dissociation constants at 25°C for Reactions (5.10b) and (5.10c) are $K_{a1} = 4.2 \times 10^{-7}$ and $K_{a2} = 5.0 \times 10^{-11}$, respectively, and the solubility of CO_2 in water is 1.0×10^{-5} M at 25°C at the current atmospheric CO_2 abundance of 360 ppmv.

(a) Calculate the concentration of $H_3O^+(aq)$, $H_2CO_3(aq)$, $HCO_3^-(aq)$, $OH^-(aq)$ and $CO_3^{2-}(aq)$ when CO_2 from the air dissolves in otherwise pure water.

(b) What is the pH of the resulting solution?

Solution. (a) From Reaction (5.10b), and the definition of acid dissociation (or equilibrium) constant for a chemical reaction,

$$\frac{[H_3O^+(aq)][HCO_3^-(aq)]}{[H_2CO_3(aq)]} = K_{a1} = 4.2 \times 10^{-7} \tag{5.12}$$

Similarly, for Reaction (5.10c)

$$\frac{[H_3O^+(aq)][CO_3^{2-}(aq)]}{[HCO_3^-(aq)]} = K_{a2} = 5.0 \times 10^{-11} \tag{5.13}$$

Also, from the definition of the ion-product constant for water at 25°C,

$$[H_3O^+(aq)][OH^-(aq)] = K_w = 1.0 \times 10^{-14} \tag{5.14}$$

If we consider 1 liter of water, 10^{-5} moles of CO_2 will dissolve in it to form 10^{-5} moles of H_2CO_3, which is distributed among $H_2CO_3(aq)$, $HCO_3^-(aq)$ and $CO_3^{2-}(aq)$. Therefore, from mass balance considerations,

$$\begin{aligned}[H_2CO_3]_{\text{initial}} &= 1.0 \times 10^{-5} \\ &= [H_2CO_3(aq)] + [HCO_3^-(aq)] + [CO_3^{2-}(aq)]\end{aligned} \tag{5.15}$$

Finally, since the equilibrium solution can have no net electric charge,

$$[H_3O^+(aq)] = [HCO_3^-(aq)] + 2[CO_3^{2-}(aq)] + [OH^-(aq)] \tag{5.16}$$

where the coefficient 2 allows for the two units of negative charge on each $CO_3^{2-}(aq)$ ion.

We now have five equations, (5.12)–(5.16), for the five unknown species concentrations. Therefore, an accurate solution to the problem could be obtained. However, the solution is simplified if we make some approximations.[3] Since $K_{a1} >> K_{a2}$, the contribution of $[H_3O^+(aq)]$ from Reaction (5.10c) is negligible compared to that from Reaction (5.10b). Also, since the only source of $CO_3^{2-}(aq)$ is from Reaction (5.10c), $[CO_3^{2-}(aq)]$ will be small compared to $[H_2CO_3(aq)]$ and $[HCO_3^-(aq)]$. Finally, since $OH^-(aq)$ derives only from the dissociation of water and an acid has been added to the water, we can assume that $[H_3O^+(aq)] >> [OH^-(aq)]$. Hence, from Eq. (5.16), $[H_3O^+(aq)] \simeq [HCO_3^-(aq)]$. Consequently, Eqs. (5.12) and (5.15) become, respectively,

$$\frac{[H_3O^+(aq)]^2}{[H_2CO^3(aq)]} \simeq 4.2 \times 10^{-7}$$

and

$$1.0 \times 10^{-5} \simeq [H_2CO_3(aq)] + [H_3O^+(aq)]$$

We now have two equations for two unknowns, which yield $[H_3O^+(aq)] \simeq 1.8 \times 10^{-6}$ M and $H_2CO_3(aq)] \simeq 8.1 \times 10^{-6}$ M. Therefore, $[HCO_3^-(aq)] \simeq [H_3O^+(aq)]$ 1.8 × 10^{-6} M. Substituting $[H_3O^+(aq)] \simeq 1.8 \times 10^{-6}$ M into Eq. (5.14) yields $[OH(aq)] \simeq 5.6 \times 10^{-9}$ M. Finally, substituting values into Eq. (5.13) gives $[CO_3^{2-}(aq)] \simeq 5 \times 10^{-11}$ M.[4]

(b) The pH of a solution is defined as

$$\text{pH} = -\log[H_3O^+(aq)]$$

Therefore, since $[H_3O^+(aq)] \simeq 1.8 \times 10^{-6}$ M,

$$\text{pH} \simeq -\log(1.8 \times 10^{-6}) = 5.7$$

Therefore, the pH of pure water exposed to an atmospheric concentration of 360 ppmv of CO_2 at 25°C is 5.7.

c. Oceanic

Most of the gases that pass from the oceans to the air originate from biological processes, which are discussed in subsection (a). Also, as we saw in the preceding subsection, the oceans can be involved in the cycling of gases between the solid Earth and the atmosphere.

The oceans are a huge reservoir for those gases that are appreciably soluble in water. For example, the atmospheric and biospheric reservoirs of CO_2 are very small (1% to 2%) compared to dissolved CO_2 in the oceans. Therefore, the oceans can serve both as a source and sink for trace gases in the atmosphere.

d. In situ *formation in the atmosphere*

Chemical reactions in the atmosphere are a major source and sink of many trace constituents. The trace gases emitted from the biosphere, solid Earth, and oceans are generally in a reduced (i.e., lower) oxidation state (e.g., C, N, and S), but when they are returned to the Earth's surface they are generally in a higher oxidation state. The processes that produce these transformations can be divided into *homogeneous gas phase*, *homogeneous aqueous phase*, and *heterogeneous reactions*.[5] These processes play important roles in atmospheric chemistry, some of which are discussed in the following sections.

5.2 Transformations by homogeneous gas-phase reactions

a. The hydroxyl radical and the nitrate radical

Most homogeneous gas-phase reactions are initiated by the absorption of solar UV radiation, that is, by a photochemical reaction. One of the most important and reactive species in the troposphere that is formed by a photochemical reaction is the hydroxyl (OH) free radical, even thought it is present in only a few tenths of a pptv at midday (or $\sim 10^6$ OH molecules per cm^3). Because it is so reactive, the lifetime of OH in the atmosphere is only about one second.[6]

Hydroxyl radicals are produced when UV radiation from the Sun decomposes O_3 into molecular oxygen and energetically excited oxygen atoms (O^*)

$$O_3 + h\nu \rightarrow O_2 + O^* \tag{5.17a}$$

where $\lambda < 0.315\,\mu m$. Most of the O^* atoms produced by Reaction (5.17a) dissipate their excess energy as heat and eventually recombine with O_2 to form O_3, which is a *null cycle* (i.e., it has no net chemical effects). However, a small fraction (~ 0.01) of the O^* atoms reacts with water vapor to form two hydroxyl radicals

$$O^* + H_2O \rightarrow OH + OH \quad (5.17b)$$

The net effect for those O* atoms involved in Reactions (5.17) is

$$O_3 + H_2O + h\nu \rightarrow O_2 + 2OH \quad (5.18)$$

Once formed, the OH radical is a powerful oxidant that reacts quickly with a number of atmospheric pollutants: CO to form CO_2, NO_2 to form HNO_3, H_2S to form SO_2, SO_2 to form H_2SO_4, CH_2O to form CO, and so on. The pivotal role of OH in transforming a large number of tropospheric gases (many of which are major pollutants) into their oxidized forms is depicted schematically in Figure 5.2. Because of this role, OH has been referred to as the "atmosphere's detergent."

Since OH is produced primarily by photochemical reactions and has a very short lifetime, it is present in the atmosphere at currently measurable levels only during the day. At night, the nitrate radical (NO_3)

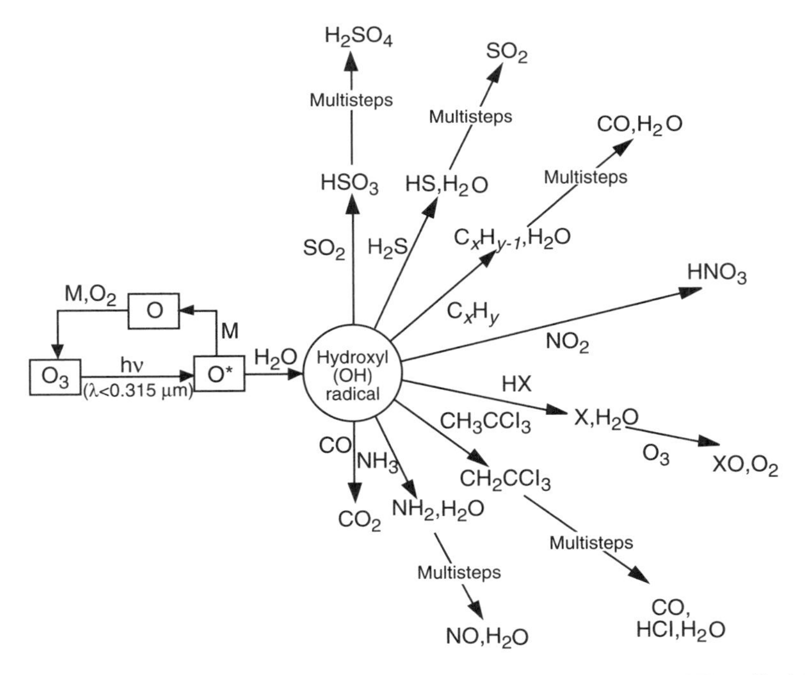

Figure 5.2. Schematic diagram to illustrate the central role of the OH radical in the oxidation of tropospheric trace gases. [Adapted with permission from *Global Tropospheric Chemistry*. Copyright © 1984 by the National Academy of Sciences. Courtesy of the National Academy Press.]

takes over from OH as the major reactive oxidant in the troposphere. It is formed by the reaction

$$NO_2 + O_3 \rightarrow NO_3 + O_2$$

During the day, NO_3 is rapidly photolyzed by solar radiation

$$NO_3 + h\nu \rightarrow NO + O_2$$

and

$$NO_3 + h\nu \rightarrow NO_2 + O$$

Although NO_3 is much less reactive than OH, at night it is present in higher concentrations than OH is during the day. Note that the NO_3 radical (often written $NO_3^{\bullet}$ to indicate an unpaired reactive electron) is a totally different species than the relatively unreactive nitrate ion in aqueous solution NO_3^- in which all electrons are paired.

Exercise 5.3. In unpolluted regions of the troposphere, OH reacts overwhelmingly with CO (~70%) and CH_4 (~30%)

$$OH + CO \rightarrow CO_2 + H \quad (5.19)$$

$$OH + CH_4 \rightarrow CH_3 + H_2O \quad (5.20)$$

What changes in the oxidation number of C are produced by Reactions (5.19) and (5.20)?

Solution. The oxidation number of the oxygen atom in CO is –2; the oxidation number of C in CO is +2. In CO_2 the oxidation number of C is +4. Hence, Reaction (5.19) increases the oxidation number (or valence state) of C by 2. The oxidation number of H in CH_4 is +1, the oxidation number of C in CH_4 is –4. In CH_3 the oxidation number of C is –3. Therefore, Reaction (5.20) increases the oxidation number of C by 1.

b. *Ozone*

Ozone plays a key but dichotomous role in the atmosphere. In both the troposphere and the stratosphere it is a key reactant.[7] In the troposphere it is a potent pollutant, but it is also the source of the "detergent" OH. In the stratosphere ozone absorbs dangerous UV radiation, thereby protecting life on Earth (see Section 10.1).

The principal source of O_3 is in the stratosphere is the photochemical reaction

$$O_2 + h\nu \rightarrow O + O \qquad (5.21a)$$

which occurs at wavelengths below 0.24 μm (these short-wavelength photons do not penetrate lower than about 30 km in the atmosphere). Reaction (5.21a) liberates atomic oxygen, which is very reactive. One path for these atoms is

$$2O + 2O_2 + M \rightarrow 2O_3 + M \qquad (5.21b)$$

for a net reaction

$$3O_2 + h\nu + M \rightarrow 2O_3 + M \qquad (5.22)$$

where M represents any other molecule which, while not directly involved in the chemical reaction, can carry away excess energy formed in the reaction. Reflecting the importance of the stratosphere as an O_3 source, O_3 concentrations increase from ~10 to 100 ppbv near the Earth's surface to peak values of ~10 ppmv between ~15 and 30 km in altitude. Photolysis of NO_2 at wavelengths below 0.38 μm is the major way in which O_3 is formed *in situ* in the troposphere

$$NO_2 + h\nu \rightarrow O + NO \qquad (5.23a)$$

$$O + O_2 + M \rightarrow O_3 + M \qquad (5.23b)$$

$$\text{Net:} \quad NO_2 + O_2 + h\nu \rightarrow NO + O_3 \qquad (5.24)$$

Reaction (5.18) is a sink for O_3 in the troposphere. Additional sinks are

$$O_3 + OH \rightarrow O_2 + HO_2 \qquad (5.25)$$

$$HO_2 + O_3 \rightarrow 2O_2 + OH \qquad (5.26)$$

Reaction (5.26) is generally the more important, because OH is converted into HO_2 (the hydroperoxyl radical) primarily by reacting with CH_4 and CO.

Although tropospheric ozone accounts for only about 10% of the total mass of O_3 in the atmosphere, most of the primary oxidation chains in the unpolluted troposphere are initiated by O_3 through its production of the OH radical (see Fig. 5.2). For example, OH from Reaction (5.18) and the H from Reaction (5.19) can form HO_2

$$H + O_2 + M \rightarrow HO_2 + M \qquad (5.27)$$

and the CH_3 from Reaction (5.20) can form the methylperoxyl radical (CH_3O_2)

$$CH_3 + O_2 + M \rightarrow CH_3O_2 + M \quad (5.28)$$

Then, in the unpolluted troposphere,

$$2HO_2 \rightarrow H_2O_2 + O_2 \quad (5.29)$$

and

$$CH_3O_2 + HO_2 \rightarrow CH_3OOH + O_2 \quad (5.30)$$

where H_2O_2 (hydrogen peroxide) is another powerful oxidant, which readily dissolves in cloud water where it can oxidize absorbed SO_2 into H_2SO_4. Methyl hydroperoxide (CH_3OOH) is also dissolved by cloud water, although there is probably not enough of it to be a major oxidant in the aqueous phase.

In the presence of NO_x, the peroxyl radicals from Reactions (5.27) and (5.28) can follow quite different paths from those described earlier. For HO_2 the path is

$$HO_2 + NO \rightarrow OH + NO_2 \quad (5.31)$$

which regenerates OH and oxidizes NO to NO_2. For CH_3O_2 the path is

$$CH_3O_2 + NO \rightarrow CH_3O + NO_2 \quad (5.32a)$$

$$CH_3O + O_2 \rightarrow HCHO + HO_2 \quad (5.32b)$$

which also oxidizes NO to NO_2. The NO_2 can then serve in Reaction (5.24) to provide the *in situ* source of tropospheric ozone. In addition, Reactions (5.32) release HO_2, which can react with NO to regenerate OH by Reaction (5.31), and then generate formaldehyde (HCHO). Formaldehyde is very reactive photochemically

$$HCHO + h\nu \rightarrow H + HCO \quad (5.33)$$

The H from Reaction (5.33) can feed back into Reaction (5.27), and the HCO can react with O_2 to produce both CO and HO_2

$$HCO + O_2 \rightarrow CO + HO_2 \quad (5.34)$$

The CO from Reaction (5.34) can participate in Reaction (5.19), and the HO_2 can produce OH and NO_2 through Reaction (5.31).

Clearly, NO_x plays a pivotal role in tropospheric chemistry by providing a path for the oxidation of methane, by oxidizing NO to NO_2, and by generating formaldehyde (see Exercise 19 in Appendix I).

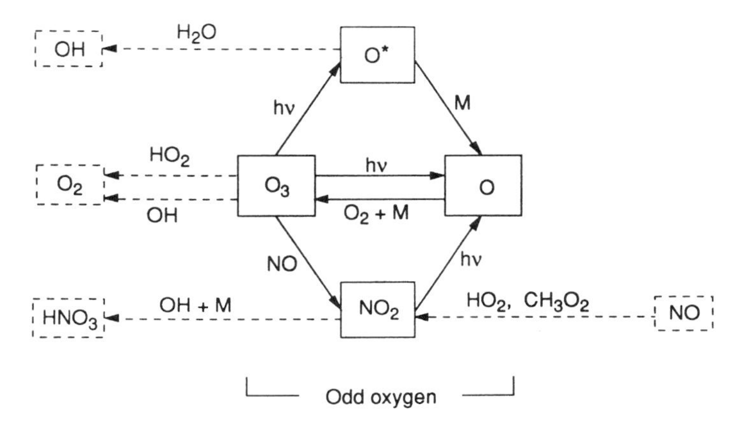

Figure 5.3. Schematic diagram showing major chemical reactions affecting odd oxygen (O_3, O, O*, NO_2) in the troposphere. Air molecules acting as third bodies are denoted by "M" (e.g., N_2, O_2, Ar, H_2O). [Adapted from J. A. Logan et al., *J. Geophys. Res.*, **86**, 7210 (1981). Copyright by the American Geophysical Union.]

c. Chemical families

It is convenient in atmospheric chemistry to identify closely related species, referred to as *chemical families*. In subsections (a) and (b) we discuss the members of one such family, comprising O, O*, and O_3, which is called the *odd oxygen* family ("odd" meaning an odd number of electrons available for bonding with another chemical species, which tends to make the atom or molecule reactive). The relationships between the members of this family are summarized in Figure 5.3. Other chemical families are NO_x (= NO, NO_2, and NO_3), NO_y, or *odd nitrogen* (which is the sum of NO_x and all oxidized nitrogen species in the air that are sources and sinks of NO_x on relatively short time scales, such as N_2O_5, HNO_3, and PAN), *odd hydrogen* (HO_x, where $x = 0, 1, 2$), and *odd chlorine* (ClO_x, where $x = 0, 1, 2$).

d. Summary

Clearly, the number of homogeneous, gas-phase chemical reactions that can occur in the atmosphere is very large. In this section we have discussed a few of these reactions. However, the reactions we have discussed are among the most important, since O_3, OH, HO_2, and H_2O_2 essentially

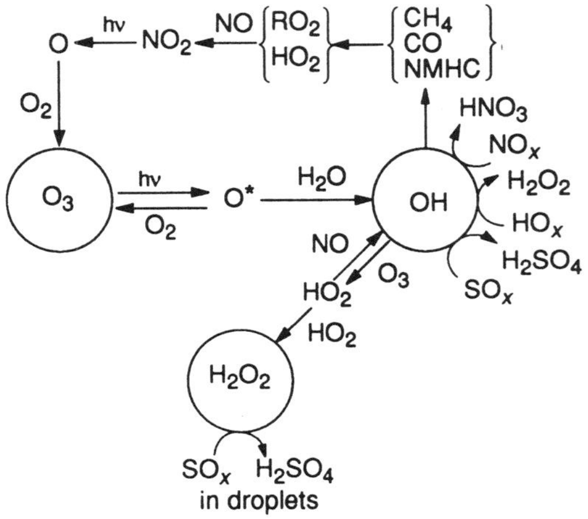

Figure 5.4. Schematic diagram illustrating the roles of the oxidants O_3, OH, HO_2 and H_2O_2 in atmospheric photochemical reactions. [Reprinted with permission from A. M. Thompson, *Science*, **256**, 1157 (1992). Copyright © 1992 American Association for the Advancement of Science.]

determine the oxidizing capacity of the atmosphere and are therefore involved in many chemical reactions (Fig. 5.4).

Four additional points, illustrated by the preceding discussion, should be emphasized:

- The importance of photochemical reactions.
- The key role that free radicals play in atmospheric chemistry.
- The importance of chemical reaction cycles that anchor the steady-state concentrations of crucial chemical species, while "spinning off" a large number of other species that can themselves be involved in additional reactions (Fig. 5.4).
- The concept of closely related chemical families of species.

5.3 Transformations by other processes

Until recently, homogeneous gas-phase reactions were thought to be the only reactions of any significance in the atmosphere. However, it is now realized that homogeneous aqueous-phase reactions in cloud droplets and heterogeneous reactions on the surfaces of aerosols and on ice particles in clouds can play important roles in atmospheric chemistry. Interestingly, this realization was prompted by two serious environmental problems that have become apparent in the latter part of the twentieth century: acid rain and the stratospheric "ozone hole" over Antarctica. We will postpone discussion of the transformation of chemical species by homogeneous aqueous-phase reactions and heterogeneous reactions until we discuss cloud chemistry (in Chapter 7) and the stratospheric ozone hole (in Section 10.2).

An interesting process for transforming nitrogen species in the atmosphere involves lightning. This is because the reaction

$$N_2 + O_2 \rightarrow 2NO \tag{5.35}$$

requires much higher temperatures than normally found in the atmosphere. However, the required temperatures are produced by the shock wave in a lightning stroke. It is estimated that $\sim 10^{26}$ molecules of NO are produced by a single lightning stroke. Since there are $\sim 50\,s^{-1}$ lightning strokes worldwide, the production rate from this source is thought to be ~5 Tg(N) per year, although there is a large uncertainty in this estimate. This production rate is a small fraction of the global nitrogen fixation, which is now dominated by human activity (see Section 8.2).

5.4 Transport and distributions of chemicals

In the planetary boundary layer (PBL), which extends from the surface of the Earth up to a height of ~1 km (depending on location and season), the atmosphere interacts directly with the Earth's surface through turbulent mixing. During the day over land, the PBL, and the chemicals in it, are generally well mixed by convection up to a height of ~1 km; at night, mixing is less efficient, and the depth of the PBL may be only a few hundred meters. Consequently, the concentrations of some chemicals in the PBL over land tend to be greater at night than during the day. Over the oceans, the diurnal cycle is much less apparent. Since chemicals in the PBL are repeatedly brought into contact with the Earth's surface, some compounds can be removed quite rapidly by deposition (see Section 5.5).

If a chemical that is emitted from the surface of the Earth is not returned to the surface or transformed in the PBL, it will eventually pass into the region above the PBL, which is referred to as the *free troposphere*. Once it is in the free troposphere, a chemical with a long residence time will follow the global circulation pattern. For example, in midlatitudes, where the winds are generally from west to east and have speeds of ~ 10 to $30\,m\,s^{-1}$, such chemicals will be distributed fairly uniformly longitudinally around the globe within 1 to 2 months. However, in the north-south direction, where wind speeds are generally much less, the distribution of chemicals will reflect more the latitudinal distribution of their sources. Since the transport of tropospheric air across the equator is relatively restricted, so is the transport of chemicals. The main effect of this is that the chemistry of the troposphere in the northern hemisphere is affected by anthropogenic emissions much more than the southern hemisphere. The chemistry of the southern hemisphere reflects more the effects of emissions from the oceans. Transport is also restricted from the free troposphere into the stratosphere; most of the upward transport is in the tropics, and the downward transport is in midlatitudes. Nevertheless, we will see in Section 10.2 that certain long-lived chemicals of anthropogenic origin can accumulate in the stratosphere, where they can have major effects.

5.5 Sinks of chemicals

The final stage in the life history of a chemical species in the atmosphere is its removal by chemical or physical processes. Chemical sinks involve transformations into other chemical species (see Section 5.3) and gas-to-particle (g-to-p) conversion, which can involve chemical and physical processes. The other important removal process for both gases and aerosols is deposition onto the Earth's surface. Deposition is of two types: wet and dry. *Wet deposition* involves the scavenging of gases and particles in the air by clouds and precipitation; it is one of the major mechanisms by which the atmosphere is cleansed (see Chapter 7).

Dry deposition involves the direct collection of gases and particles in the air by vegetation, the Earth's surface, and the oceans. It is much slower than wet deposition, but it acts continuously rather than episodically. The dry deposition of a gas can be quantified empirically by defining its *deposition velocity* for a particular surface:

$$\text{deposition velocity} = \frac{\text{flux of the substance to the surface}}{\begin{array}{c}\text{mean concentration of}\\ \text{the substance near the surface}\end{array}}$$

where the concentration is usually measured at a height of 1 m above the surface. For example, the deposition velocity of CO and SO_2 onto soils and vegetation are ~0.05 and 0.14 to 2.2 $cm\,s^{-1}$, respectively. The reciprocal of the deposition velocity is called the *resistance* (of the surface to uptake). The total resistance is given by the sum of the resistances for the various transport processes. For example, in the uptake of SO_2 by vegetation, the total resistance is equal to the sum of the resistances to transfer of SO_2 through the air, through the boundary layer, and through the surface layers of the plant itself.

The oceans are important sinks for many trace gases. For example, the global fluxes of O_3 and SO_2 to the oceans are about 100 and $5\,Tg\,a^{-1}$, respectively. The fluxes depend on how undersaturated the oceans are with respect to the gas. If the surface layers of the ocean are supersaturated with a gas, then the flux is from the ocean to the atmosphere (e.g., the estimated global flux of DMS from the ocean to the atmosphere is $\sim 25\,Tg(S)\,a^{-1}$, which is due to the fact that marine algae produce DMS in the surface waters).

Exercises

See Exercises 1(o)–(s) and Exercises 18–26 in Appendix I.

Notes

1 Note that some of the more exotic compounds in the atmosphere originate from microscopic organisms.
2 The CO_2 and H_2O in Reaction (5.8a) combine to form carbonic acid (H_2CO_3); see Exercise 5.2.
3 Approximations, based on a knowledge of the chemistry involved, are common when dealing with atmospheric or geochemical problems where great accuracy is generally unnecessary.
4 In calculations such as this, where several approximations are made, the solutions should be checked by substituting the derived values back into the original equations to see if reasonable equalities are obtained.
5 Chemists define a *homogeneous* reaction as one in which all of the reactants are in the same phase, and a *heterogeneous* reaction as one involving reactants in two or more phases. Unfortunately, these definitions differ from those used in cloud physics, where a homogeneous process is defined as one involving just one substance (in any phase) and a heterogeneous process as one involving more than one substance. In this book we use the chemical definitions.
6 Because of the extremely low concentration and short lifetime of OH it is very difficult to measure. Even so, using modern techniques, the atmospheric concentration of OH during the day in the Earth's atmosphere has been determined to within about a factor of two.
7 One reason ozone is very reactive is because it reacts exothermically (i.e., releases heat) with many substances and the rates of these reactions are quite high. The O_2–O bond energy is only $105\,kJ\,mol^{-1}$, whereas typical bond energies are three times this value.

6
Atmospheric aerosols

Apart from cloud and precipitation particles, which have relatively large sizes, the Earth's atmosphere consists a mixture of gases, small solid particles, and small liquid droplets. Mixtures of air with small solid particles and small droplets are called *aerosols*. The small particles and droplets themselves (called *aerosol particles*, but often loosely referred to as simply aerosols) are important not only in air chemistry but in determining visibility, the formation of cloud particles, atmospheric radiation, and atmospheric electricity. In this chapter we will be concerned primarily with chemical aspects of atmospheric aerosols, but we will start by describing two of their important physical attributes, namely, their concentrations and size distributions.

6.1 Aerosol concentrations and size distributions

a. Total number and mass concentrations

One of the oldest and most convenient techniques (which in various modified forms is still in use) for counting the number concentrations of atmospheric aerosols is the Aitken[1] nucleus counter. In this instrument, humid air is expanded rapidly so that it cools and becomes supersaturated by several hundred percent with respect to water. At these high supersaturations, water condenses onto virtually all of the particles in the air to form a cloud of small water droplets in the chamber of the counter. The number concentration of droplets in this cloud (which is close to the total number concentration of aerosol particles) can be determined automatically by optical techniques or by allowing the droplets to settle onto a substrate where they can be counted. The concentration of aerosol particles measured with an Aitken nucleus counter is referred to as the *Aitken* (or *condensation*) *nucleus count* (*CN count* for short).

Aitken nucleus counts near the Earth's surface vary widely among different locations, and they can fluctuate in time by more than an order of magnitude at any one site. Generally, their average values near the Earth's surface are ~$10^3\,cm^{-3}$ over the oceans, ~$10^4\,cm^{-3}$ over rural land areas, and ~$10^5\,cm^{-3}$ or higher in polluted air.

Shown in Figure 6.1 are curves representing smoothed fits to (quite often variable) measurements of the vertical distributions of CN counts in various locations. The CN counts for aerosols in remote (i.e., far removed from local sources of pollution) continental areas generally decrease with increasing height. Aitken nucleus counts over the oceans (marine) show fairly constant values in the troposphere above ~2 km; at lower levels, the CN count is sometimes less and sometimes greater than aloft. In polar regions above ~1 km the CN counts are also often fairly constant with height at ~$200\,cm^{-3}$, with lower values below ~1 km. These results suggest that above the PBL, the "natural background" CN count in the troposphere may be ~$300\,cm^{-3}$. However, in many locations, anthropogenic pollution, biogenic emissions, or biomass burning often increases this value considerably.

Smooth fits to measurements of the vertical distributions of the mass concentrations of atmospheric aerosols are shown in Figure 6.2. Mass concentrations near the surface vary from average values of ~$3\,\mu g\,m^{-3}$ in polar regions to ~$100\,\mu g\,m^{-3}$ in deserts. The mass concentrations generally decrease with height. For remote continental and marine regions, the natural background for the mass concentrations is ~$1\,\mu g\,m^{-3}$ above ~2 km.

b. Aerosol size spectra

The *size spectra* (or *size distributions*) of aerosols refer to their number, surface area, or volume concentrations as a function of aerosol diameter. The *size distribution function*, $f(D)$, for the number concentrations of aerosols is defined by

$$f(D) = \frac{1}{N}\frac{dN}{dD} \tag{6.1}$$

where N is the total number concentrations of aerosols, and dN the number concentrations of aerosols with diameters between D and $D + dD$.[2] Aerosol surface area (S) and volume (V) distribution functions may

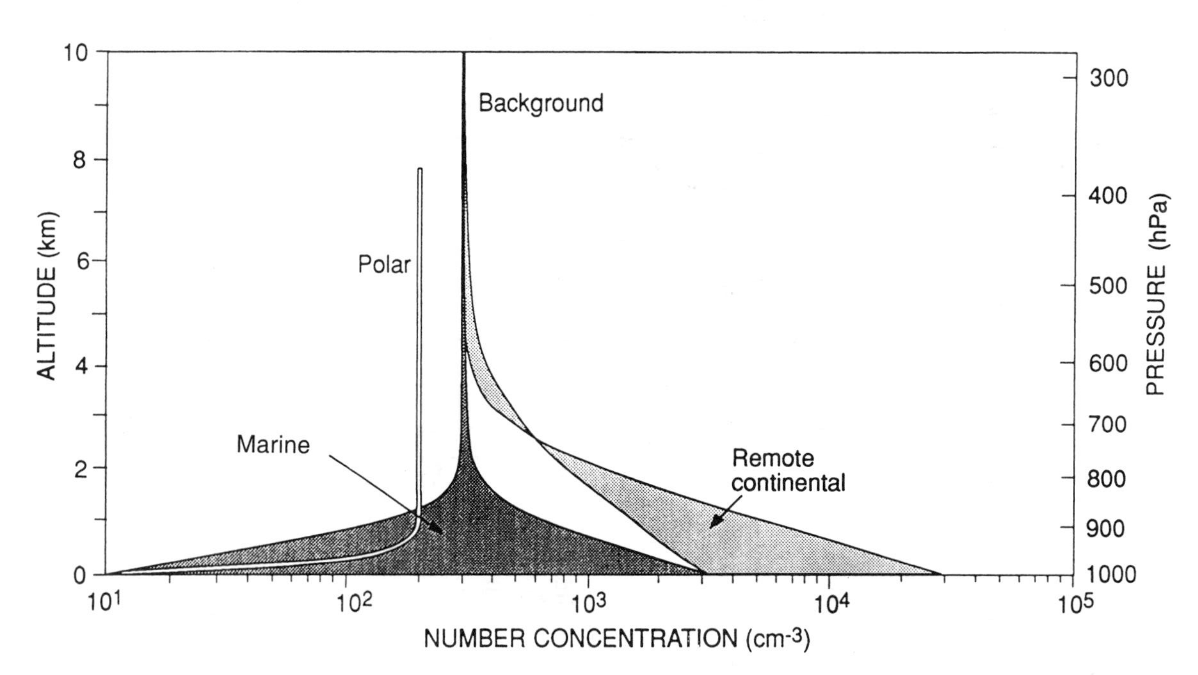

Figure 6.1. Vertical profiles of aerosol number concentrations. The ranges of concentrations are shown for marine and remote continental air. [From R. Jaenicke in *Aerosol-Cloud-Climate Interactions*, Ed. P. V. Hobbs, Academic Press, p. 23 (1993).]

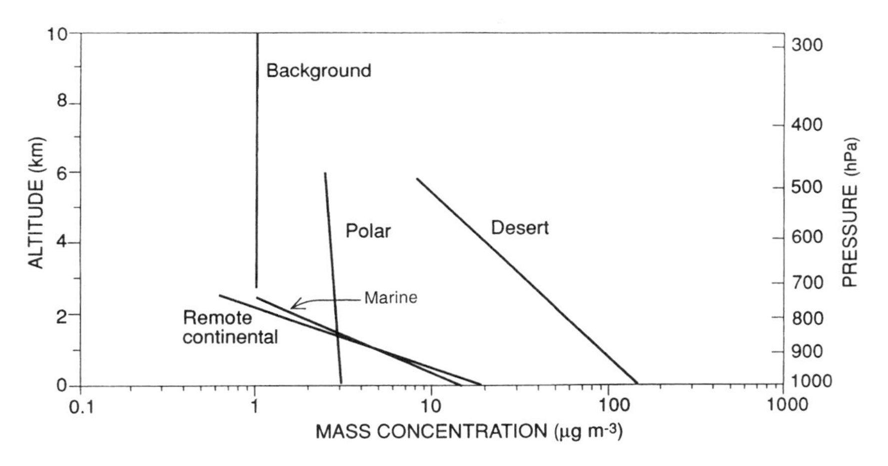

Figure 6.2. Vertical profiles of aerosol mass concentrations. [From R. Jaenicke in *Aerosol-Cloud-Climate Interactions*, Ed. P. V. Hobbs, Academic Press, p. 22 (1993).]

be defined in an analogous way to Eq. (6.1). Measurements on atmospheric aerosols with D $\gtrsim 0.2\,\mu$m can often be represented by

$$\frac{dN}{dD} = C_1 D^{-(\beta+1)} \tag{6.2}$$

where C_1 and β are constants. Since

$$d(\log D) = \frac{1}{2.302} d(\ln D) = \frac{1}{2.302} \frac{dD}{D}$$

Eq. (6.2) can be written as

$$\frac{dN}{d(\log D)} = C_2 D^{-\beta} \tag{6.3}$$

where $C_2 = 2.302\, C_1$.

Some measurements of aerosol number distributions in urban (polluted) air are shown in Figure 6.3. In this plot the ordinate represents the left side of Eq. (6.3), the abscissa is D, and both are plotted on a logarithmic scale. Therefore, the slope of a straight line on such a plot is equal to $-\beta$. For particles with $D \gtrsim 0.2\,\mu$m, the data shown in Figure 6.3 follow a straight line with slope of about –3. Hence, the number concentrations of these particles follow Eqs. (6.2) or (6.3) with $\beta = 3$; for this particular value of β, Eq. (6.3) is called the *Junge*[3] *distribution*.

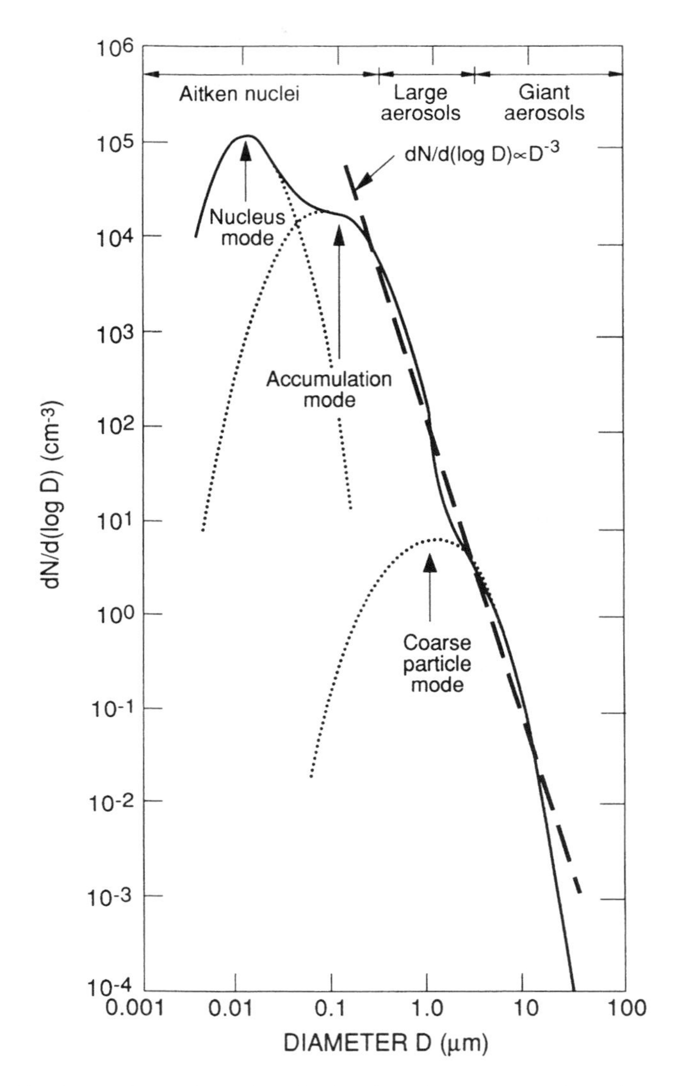

Figure 6.3. Aerosol number size distributions in urban air. The heavy dashed line shows a Junge distribution fit to the data for the large and giant aerosols. [Adapted from K. T. Whitby, *Atmos. Environ.*, **12**, 135 (1978). Copyright © 1978 with permission from Elsevier Science.]

Exercise 6.1. Show that for the Junge distribution the aerosols contained within each logarithmic increment of diameter contribute equally to the total volume concentration of the aerosols.

Solution. In Eq. (6.3), dN is the number concentration of aerosols within the size interval $d(\log D)$. The volume concentration dV of aerosols in this interval is proportional to $D^3 dN$. For a Junge distribution, $\beta = 3$ in Eq. (6.3), therefore, $D^3 dN \propto d(\log D)$. Hence, $D^3 dN/d(\log D)$, or $dV/d(\log D)$, is a constant.

It follows from Exercise 6.1 that although aerosols with diameters from, say, 0.2 to 2 μm (*large aerosols*) are present in much higher number concentrations than aerosols with diameters from 2 to 20 μm (*giant aerosols*), for a Junge distribution (the heavy dashed line in Fig. 6.3) the large and giant aerosols contribute equally to the total volume of aerosols (or to the total mass of the aerosol, if the aerosol density is independent of D). However, since the number concentration of aerosols with diameter from about 0.2 to 0.02 μm does not increase in concentration with decrease in particle diameter as rapidly as the Junge distribution (i.e., they fall below the dashed line in Fig. 6.3), these small particles contribute only ~10% to 20% to the total volume of aerosols. Nevertheless, aerosols with $D \lesssim 0.2\,\mu$m dominate the total number concentration (N), that is, the Aitken nucleus count; for this reason these particles are referred to as *Aitken nuclei*.

It can be seen from Figure 6.3 that the complete aerosol number distribution in urban air is composed of three modes: the *nucleus* (or *nucleation*) *mode*, which peaks at $D \simeq 0.01\,\mu$m, the *accumulation mode*, which peaks at $D \simeq 0.1\,\mu$m, and the *coarse particle mode*, which peaks at $D \simeq 1\,\mu$m. The nucleation mode is produced by the condensation of gases (particularly H_2SO_4), and is therefore prominent close to sources of pollution. The accumulation mode is due to the coagulation of smaller particles, the condensation of gases onto existing particles, and from the particles left behind when cloud drops evaporate. Consequently, a prominent accumulation mode is characteristic of an aged aerosol. The coarse particle mode is produced by wind-blown dusts, industrial processes that produce fly ash and other large aerosols, and sea salt from the oceans.

Shown in Figure 6.4 are some measurements of aerosol number distributions in rural continental and marine air. Comparisons of Figures 6.3 and 6.4 show that the main differences in the aerosol spectra for urban polluted, rural continental, and marine air is that the urban air has

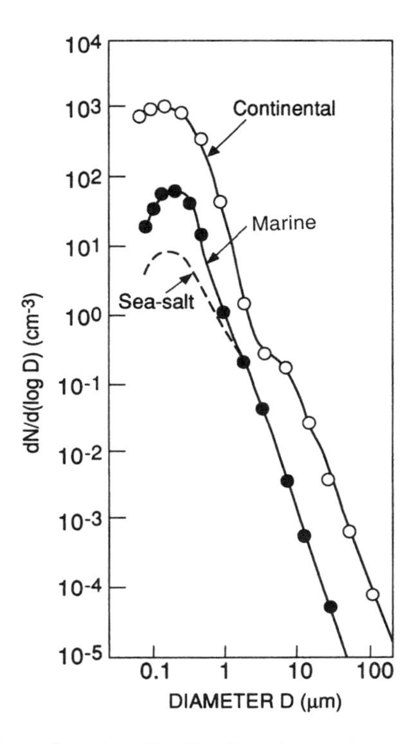

Figure 6.4. Aerosol number size distributions in continental rural air (Hungary) and in marine air in the southern hemisphere. Also shown is the typical contribution from sea-salt particles over the oceans. [Adapted from G. Götz, E. Mészáros, and G. Vali, *Atmospheric Particles and Nuclei*, Akadémiai Kiadó, Budapest, p. 39 (1991).]

higher concentrations of Aitken nuclei. For $D \gtrsim 0.2\,\mu$m all three spectra follow quite closely a Junge distribution, but with the aerosol concentrations in the marine air much less than those in the continental and urban air.

Sea-salt concentrations, measured in stormy weather over the Indian Ocean, are shown in Figure 6.4. In remote oceanic locations such as this, practically all of the giant aerosols are sea salt. However, with increasing distance inland, sea-salt concentrations fall off rapidly. Sea salt makes only a small contribution to the Aitken nuclei, even over the oceans.

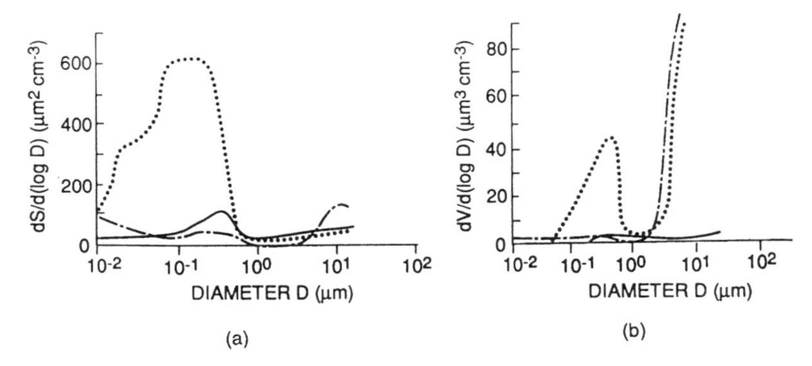

Figure 6.5. Surface (a) and volume (b) distributions based on aerosol size measurements in Denver, Colorado. Urban polluted air (. . .), continental air (——), and continental air with dust (– · – ·). [From K. Willeke and K. T. Whitby, *J. Air Poll. Cont. Assoc.*, **25**, 532 (1975).]

Since the surface area dS of aerosols in the diameter range D to $D + dD$ is equal to $\pi D^2 dN$

$$\frac{dS}{d(\log D)} = \pi D^2 \frac{dN}{d(\log D)} \tag{6.4}$$

Similarly, for the volume of particles, $dV = (\pi/6)\, D^3 dN$, and

$$\frac{dV}{d(\log D)} = \frac{\pi}{6} D^3 \frac{dN}{d(\log D)} \tag{6.5}$$

In other words, surface area and volume distribution can be obtained from the number distribution by applying the weighting factors for surface area and volume, respectively.

Shown in Figure 6.5 are some aerosol surface area and volume distributions for urban polluted and continental air. These plots are quite different in shape from aerosol number distribution plots. This is because the weighting factors in Eqs. (6.4) and (6.5) cause small fluctuations in the slope of the number distribution curve around values of –2 and –3 to be translated into local maxima and minima in the surface and volume distributions curves, respectively (see Exercise 28 in Appendix I). Consequently, the prominent maxima in the aerosol surface area and volume distributions, particularly for urban polluted air, shown in Figure 6.5 reflect the accumulation mode and the coarse particle mode, which show

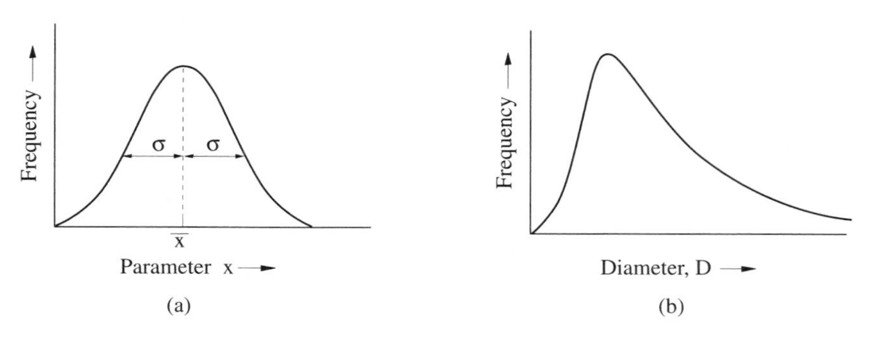

Figure 6.6. Examples of (a) a normal distribution, and (b) a log-normal distribution.

up only as changes in slope in the composite number distribution curve shown in Figure 6.3.

The sizes of atmospheric aerosol vary with relative humidity. This is because some of the particles are water soluble. The relative humidity at which a water-soluble particle starts to increase significantly in size (called the *deliquescent point*) depends primarily on the chemical composition of the particle. For example, magnesium chloride ($MgCl_2$), ammonium nitrate (NH_4NO_3), sodium chloride (NaCl), and potassium chloride (KCl) deliquesce at relative humidities of 33%, 62%, 75%, and 84%, respectively.

If the frequency with which the magnitude of a quantity x occurs is plotted graphically on linear scales and a symmetrical bell-shaped curve is obtained (as shown in Fig. 6.6a), x is said to be *normally distributed*. Such a distribution can be described by two numbers: the average (or mean) value of $x(\bar{x})$ and the standard deviation of $x(\sigma)$. The diameters (D) of atmospheric particles are generally not normally distributed, but instead are skewed (Fig. 6.6b). However, if instead of plotting the diameter D of the particle on the abscissa $\log D$ is plotted, this will tend to spread out the smaller size ranges and compress the larger ones. If the resulting curve is bell-shaped, the particles are said to be *log-normally distributed*. By analogy with the definitions of mean and standard deviation for a normal distribution, corresponding values for a log-normal distribution, referred to as the *geometric mean diameter* (D_g) and the *geometric standard deviation* (σ_g), are defined as[4]

$$\log D_g = \frac{\sum_i n_i \log D_i}{\sum_i n_i} \tag{6.6a}$$

and

$$\log \sigma_g = \left[\frac{\sum_i n_i (\log D_g - \log D_i)^2}{\left(\sum_i n_i\right) - 1} \right]^{1/2} \tag{6.6b}$$

where, n_i is the number of particles with diameter D_i.

For modeling purposes, each of the various modes of an atmospheric aerosol size distribution (as depicted, for example in Fig. 6.3) are often approximated by a log-normal distribution, which can be represented by

$$\frac{dN}{dD} = \frac{N}{(2\pi)^{1/2} D \ln \sigma_g} \exp\left[-\frac{(\ln D - \ln D_g)^2}{2 \ln^2 \sigma_g} \right] \tag{6.7}$$

where, as before, dN is the number concentrations of particles with diameters between D and $D + dD$, and N the total number concentration of particles of all sizes. It can be shown that for a log-normal distribution, D_g is the *median* diameter, that is, exactly one-half of the particles are smaller than D_g and one-half are larger than D_g; and, σ_g is the particle diameter below which ~84% of the particles lie divided by the median diameter D_g.

6.2 Sources of aerosols

a. Biological

Biological (or biogenic) aerosols are released into the atmosphere from plants and animals. They include seeds, pollen, spores, and fragments of animals and plants, which are usually between 1 and 250 μm in diameter, and bacteria, algae, protozoa, fungi, and viruses, which are generally <1 μm in diameter. Some characteristic concentrations are maximum values of grassy pollens >200 m^{-3}; fungal spores (in water) ~100 to 400 m^{-3}; bacteria over remote oceans ~0.5 m^{-3}; bacteria in New York City ~80 to 800 m^{-3}; and bacteria over sewage treatment plants ~10^{10} to 10^{11} m^{-3}. Microorganisms live on skin: when you change your clothes, you propel ~10^5 bacteria per minute into the air, with diameters ranging from ~1 to 5 μm!

The oceans are an important source of biogenic aerosols. They are injected into the atmosphere by the bursting of air bubbles (see subsection c) and by seafoam.

Smoke from forest fires can be a large source of aerosols: soot particles (primarily organic compounds and elemental carbon) and fly ash are directly injected into the air by forest fires. Several million grams of aerosols can be released by the burning of 1 hectare (10^4 m^2). It is estimated that 200 to 450 Tg of aerosols (containing 90 to 180 Tg of elemental carbon) are released into atmosphere each year by biomass burning. The number and volume distributions of aerosols from forest fires peak at ~0.1 and 0.3 μm diameter, respectively. Some biogenic particles (e.g., bacteria from vegetation) can nucleate ice in clouds.

The biogenic fraction of the atmospheric aerosol number concentration for particles with diameter <2 μm can reach 50%; for giant particles it is ~10%.

b. Solid earth

The transfer of particles to the atmosphere from the Earth's surface is caused by winds and atmospheric turbulence. To initiate the motion of particles on the Earth's surface requires frictional speeds in excess of certain threshold values that depend on the size of particle and the type of surface. The threshold values are least (~0.2 m s^{-1}) for particles 50 to 200 μm in diameter (smaller particles adhere better to the surface) and for soils containing 50% clay or tilled soils. To achieve a frictional speed of 0.2 m s^{-1} requires a wind speed at a few meters above ground level of several meters per second. For a frictional speed (u_f) much greater that the threshold value, the horizontal flux of particles in the air through a vertical plane normal to the wind direction increases approximately as u_f^3 and the vertical flux of particles increases roughly as u_f^5. A major source of smaller (10 to 100 μm diameter) aerosol particles is a process termed *saltation*, in which larger sand grains become airborne, fly a few meters and then land on the ground creating a burst of dust particles.

On the global scale, deserts (which cover about one-third of the land surface) are the main source of atmospheric aerosols from the Earth's surface. They provide ~2,000 Tg a^{-1} of mineral aerosols.

Volcanoes inject gases (which can be converted into particles by g-to-p conversion) and particles into the atmosphere. The large particles have short residence times, but the small particles may be transported globally particularly if they are blasted to high altitudes. Volcanic emissions play an important role in stratospheric chemistry (see Section 10.3).

c. Oceans

The oceans are one of the most important sources of atmospheric aerosols (~1,000 to 5,000 $Tg\,a^{-1}$, although this includes giant particles that are not transported very far). The major mechanism for ejecting ocean materials into the air is bubble bursting (some materials enter the air in drops torn from windblown spray and foam but, since these drops are large, their residence times in the air are very short).

Giant aerosols composed of sea salt originate from drops ejected into the air when air bubbles burst at the ocean surface. From 1 to 5 such drops break away from each jet that forms when a bubble bursts (Fig. 6.7c), and these jet drops are thrown about 15 cm up into the air. Some of these drops subsequently evaporate and leave behind sea-salt particles with diameters >2 μm. Much smaller droplets are produced when the upper portions of air bubble films burst at the ocean surface; these are called film droplets (Fig. 6.7b). Bubbles ≳2 mm in diameter each eject between 100 and 200 film droplets into the air. After evaporation, the film droplets leave behind sea-salt particles with diameters less than ~0.3 μm. The average rate of production of sea-salt particles over the oceans is estimated to be ~100 $cm^{-2}s^{-1}$. Hygroscopic salts (NaCl (85%), KCl, $CaSO_4$, $(NH_4)_2SO_4$) account for ~3.5% of the mass

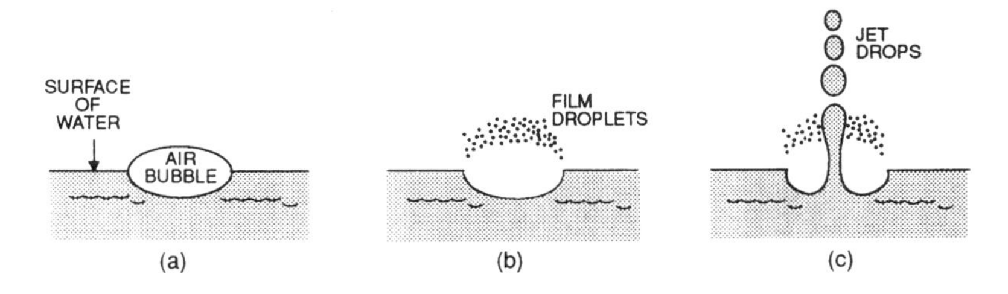

Figure 6.7. Schematic diagrams to illustrate the manner in which film droplets and jet drops are produced when an air bubble bursts at the surface of water. Over the oceans, some of the droplets and drops evaporate to leave sea-salt particles and other materials in the air. The time between (a) and (b) is ~2 ms. The size of the jet drops are ~15% of the diameter of the air bubble; the film droplets are ~5 to 30 μm diameter before evaporation.

of seawater. These materials are injected into the atmosphere by bubble bursting over the oceans. In addition, organic compounds and bacteria in the surface layers of the ocean are transported to the air by bubble bursting.

d. Anthropogenic

The global input of aerosols into the atmosphere from anthropogenic activities is about 20% (by mass) of that from natural sources. The main anthropogenic sources of aerosols are dust from roads, wind erosion of tilled land, fuel combustion, and industrial processes. For particles with diameters >5 μm, direct emissions from anthropogenic sources dominate over aerosols that form in the atmosphere by g-to-p conversion of anthropogenic gases. However, the reverse is the case for smaller particles, where g-to-p conversion is the overwhelming source of anthropogenically derived aerosols. This is why Aitken nuclei are far more numerous in urban polluted air than in continental or marine air (see Figs. 6.3 and 6.4).

In 1997, total worldwide anthropogenic direct emissions of aerosol <10 μm diameter were estimated to be ~350 Tg a^{-1} (excluding secondary SO_4^{2-}, O_3^- and organics). About 35% of the aerosols in the atmosphere were from airborne sulfate produced by the oxidation of SO_2 emissions. Aerosol emissions worldwide were dominated by fossil fuel combustion (primarily coal) and biomass burning. In 1997 these emissions were projected to double by the year 2040, largely from fossil fuel combustion, with the greatest growth in emissions from China and India.

During the twentieth century, anthropogenic emissions of aerosols were a small fraction of those from natural sources. However, it is projected that by 2040, anthropogenic aerosol emissions could become comparable to those from natural sources.

e. In situ *formation*

In situ formations of aerosols in the atmosphere by g-to-p, particularly involving gases produced by human activities, is an important source of Aitken nuclei. Since this is a transformation process, it is discussed in the following section.

Chemical reactions in cloud droplets also produce material that is left behind as aerosols when the droplets evaporate (see Section 7.4).

6.3 Transformations of aerosols

a. Enrichment

The compositions of aerosols lofted into the air are similar to but not always the same as those of the surfaces from which they originate. For example, aerosols in marine air often contain much higher concentrations of lead and mercury than does seawater. This phenomenon is referred to as *enrichment*. The *enrichment factor EF*(X) of an element X in aerosols is defined by

$$EF(X) = \frac{[\mathrm{X}]_{\mathrm{aerosol}}}{[\mathrm{Ref}]_{\mathrm{aerosol}}} \div \frac{[\mathrm{X}]_{\mathrm{source}}}{[\mathrm{Ref}]_{\mathrm{source}}} \tag{6.8}$$

where [Ref] is the concentration of an appropriate reference element (e.g., Al if X derives from the Earth's crust, and Na if X derives from the ocean).

A possible explanation for enrichment is that the surface layers of the land or ocean, from which the aerosols primarily derive, can have different chemical compositions from the underlying (bulk) substances. Differential vaporization of chemical species, both from the Earth's surface and from suspended aerosols, can also cause enrichment.

b. Gas-to-particle conversion

Gases in the atmosphere may condense onto existing particles, thereby increasing the mass (but not the number) of aerosols. Gases may also condense to form new particles in the air. The former path is favored when the surface area of existing particles is high and the supersaturation of the gases is low. If new particles are formed, they are generally in the Aitken nucleus size range. The quantities of aerosols produced by g-to-p conversion exceed those from direct emissions in the case of anthropogenically derived aerosols, and are comparable to direct emission in the case of naturally derived aerosols.

Three major chemical species are involved in g-to-p conversion: sulfur, nitrogen, and organic and carbonaceous materials. Various sulfur gases (e.g., H_2S, CS_2, COS, DMS) can be oxidized to SO_2 (Fig. 6.8). The SO_2 is then oxidized to sulfate ($SO_4^{=}$), the dominant gas phase routes being

$$SO_2 + OH + M \rightarrow HSO_3 + M \tag{6.9a}$$

$$HSO_3 + O_3 \rightarrow HO_2 + SO_3 \tag{6.9b}$$

$$SO_3 + H_2O \rightarrow H_2SO_4 \tag{6.9c}$$

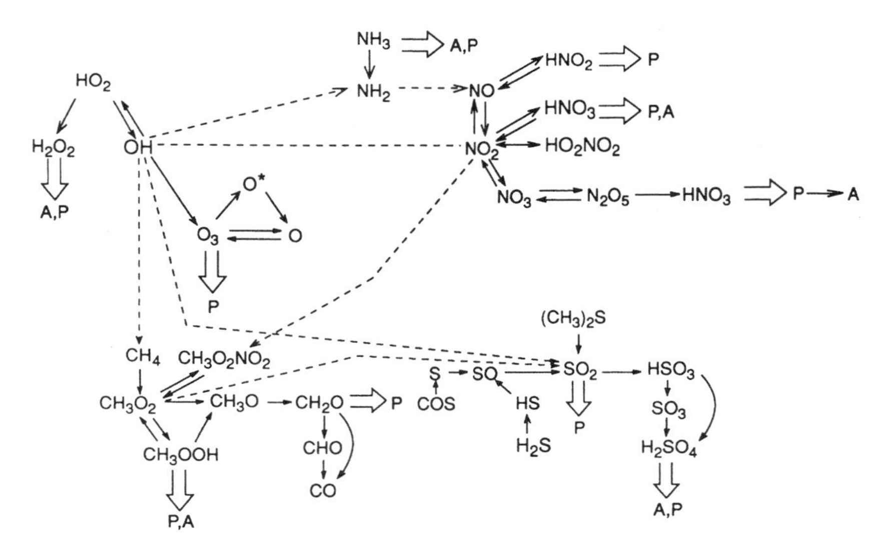

Figure 6.8. Schematic diagram of some of the primary pathways for trace gases to be converted to aerosols (A) in the troposphere. The major reactions involving gas-phase constituents are indicated by the solid lines. Interactions between chemical families are indicated by the dashed lines. Pathways leading to incorporation into precipitation (P) are also shown. [Adapted from R. P. Turco et al. in *Heterogeneous Atmospheric Chemistry*, ed. D. R. Schryer, American Geophysical Union, p. 234 (1982). Copyright © 1982 by the American Geophysical Union.]

However, on a global scale, heterogeneous reactions of SO_2 in cloud water dominate the conversion of SO_2 to $SO_4^=$ (see Section 7.4).

Over the oceans the main source of sulfates derives from DMS. Sulfates are also produced (in the Aitken nucleus size range) in the vicinity of clouds, by the combination of water molecules and sulfuric acid molecules (see Section 7.8). Sulfates are an important component of particles in the accumulation mode. Sulfates also play roles in atmospheric visibility (particularly in polluted air), as nuclei upon which cloud droplets form (see Section 7.2), in the stratosphere (see Section 10.3), and they contribute to the anthropogenic modification of climate (see Section 7.2).

Nitric acid can form from N_2O_5 in cloud water; thus, evaporation of cloud water is a source of nitrate particles (Fig. 6.8).

Organic and carbonaceous aerosols are produced by g-to-p conversion from gases released from the biosphere, and from volatile compounds of crude oil that leak to the Earth's surface. Carbonaceous particles emitted directly into the atmosphere derive mainly from biomass fires.

c. *Coagulation*

Aerosol particles with diameter $<1\,\mu$m are in constant random motion due to bombardment by gas molecules (i.e., Brownian motion). This motion causes aerosol particles to collide. If it assumed that every collision results in adherence, and that the aerosol are monodispersed (i.e., all of the same size), the rate of decrease in the concentration of the aerosols per unit volume, N, is given by

$$-\frac{dN}{dt}=\frac{4}{3}\frac{kT}{\mu}\left(1+\frac{2A\lambda}{D}\right)N^2 \tag{6.10}$$

where, k is Boltzmann's constant, T temperature (in K), μ the dynamic viscosity of air ($1.82 \times 10^{-5}\,\mathrm{N\,s\,m^{-2}}$ at 20°C), λ the mean free path of the gas molecules ($6.53 \times 10^{-8}\,\mathrm{m}$ at 20°C and 1 atm), D the diameter of the particles, and A (called the *Stokes–Cunningham correction factor*) is given by

$$A \simeq 1.257+0.4\exp\frac{-0.55D}{\lambda} \tag{6.11}$$

Coagulation shifts particles from smaller- to larger-sized categories, where removal from the atmosphere by sedimentation becomes more important (see Section 6.7). For polydispersed aerosols (i.e., particles of different sizes), different sedimentation speeds can also produce collisions as faster falling particles collide with slower moving particles.

6.4 Chemical composition of aerosols

Atmospheric aerosols may be divided into (a) water-soluble inorganic salts, (b) minerals from the Earth's crust that are insoluble in water or organic solvents, and (c) organics, some of which are soluble in water and others insoluble. Percentage contributions from these three classes of aerosols for urban and rural air in Germany are listed in Table 6.1. However, these percentages are highly variable. For example, the water-soluble fraction can range from 30% to 80%. Giant particles are composed of roughly equal amounts of water-soluble compounds and insoluble minerals. The amount of water-soluble material increases with decreasing particle size at the expense of the mineral component.

Table 6.1. *Classification by solubility of continental aerosols in Europe*[a]

	Mass fraction (%)	
Component	Urban	Rural
Water-soluble inorganic salts	30	43
Insoluble minerals from crust	35	25
Water-soluble organics	28	25
Water-insoluble organics	5	6

[a] Adapted from P. Winkler, *Meteorol. Rundsch.*, **27**, 129, 1974.

In the following discussion we divide aerosols simply into inorganic and organic.

a. Inorganic aerosols

Except for marine aerosols, the mass concentrations of which are dominated by sodium chloride, sulfate is one of the prime contributors to the mass concentration of atmospheric aerosols. The mass fractions of SO_4^{2-} range from ~22% to 45% for continental aerosols to ~75% for aerosols in the Arctic and Antarctic. Since the sulfate content of the Earth's crust is too low to explain the large percentage of sulfate in aerosols, most of it must derive from g-to-p conversion of SO_2. The sulfate is contained mainly in submicrometer-diameter aerosols, with a peak in the accumulation mode (near $D = 0.6\,\mu m$).

Ammonium (NH_4^+) is the main cation associated with SO_4^{2-} in continental aerosol; it is produced by gaseous ammonia neutralizing sulfuric acid to produce ammonium sulfate

$$2NH_3(g) + H_2SO_4(g) \rightarrow (NH_4)_2SO_4(s) \tag{6.12}$$

The ratio of the molar concentrations of NH_4^+ to SO_4^{2-} ranges from ~1 to 2, corresponding to an aerosol composition intermediate between that for NH_4HSO_4 and $(NH_4)_2SO_4$.

Nitrate (NO_3^-) is also common in continental aerosols, where it extends over the diameter range ~0.2 to $20\,\mu m$ with a peak in mass concentration in the coarse-particle mode. It derives, in part, from the condensation of $HNO_3(g)$ onto larger and more alkaline mineral aerosol particles.

In marine air the main contributors to the mass concentration of aerosols are Na^+, Cl^-, Mg^{2+}, SO_4^{2-}, K^+, and Ca^{2+}. Apart from SO_4^{2-}, these compounds are mainly in the coarse-particle mode because they originate from sea salt derived from bubble bursting. Sulfate mass concentrations peak in both the coarse-particle and accumulation modes; the latter is due to g-to-p conversion of SO_2 that derives primarily from biogenic gases (e.g., DMS).

In marine air nitrate occurs in larger sized particles than sulfate, with a significant fraction in the coarse-particle mode. Since seawater contains negligible nitrate, the nitrate in these aerosols must derive from gaseous HNO_3 followed by g-to-p conversion. Since g-to-p conversion is expected to produce aerosols in the accumulation mode, this suggests that the nitrate in marine aerosols is produced by g-to-p conversion in the liquid phase (see Section 7.4).

Over the oceans, aerosols show a deficit of Cl and Br and a surplus of sulfate, ammonium, and nitrate relative to seawater. Elements from the Earth's crust are also found in oceanic aerosols, even thousands of miles from land. The composition of continental aerosols differ appreciably from crustal rock and average soils. The enrichment factors of some of the major elements (e.g., Si, Al, Fe) can differ by factors of ~3, and some minor elements (e.g., Cu, Zn, Ag) are enriched by several orders of magnitude.

b. Organic aerosols

Organic compounds form an appreciable fraction of the mass of atmospheric aerosols (see Table 6.1), The most abundant organics in urban aerosols are higher molecular weight alkanes ($C_x H_{2x+2}$), ~1,000 to 4,000 $ng\,m^{-3}$, and alkenes ($C_x H_{2x}$), ~2,000 $ng\,m^{-3}$. Many aerosols in urban smog are by-products of photochemical reactions involving hydrocarbons and nitrogen oxides, which derive from combustion. Polycyclic aromatic hydrocarbons (PAH), such as napthalenes, are of particular concern because they are carcinogens. Recent airborne studies on the East Coast of the United States show that carbonaceous materials account, on average, for about 50% of the total dry aerosol mass.

6.5 Transport of aerosols

Aerosols are transported by the airflows they encounter during the time they spend in the atmosphere. The transport can be over intercontinental, and even global, scales. Thus, Saharan dust is transported to the

Americas, and dust from the Gobi Desert can reach the west coasts of Canada and the United States. If the aerosols are produced by g-to-p conversion, long-range transport is likely. This is because the time taken for g-to-p conversion and the relatively small sizes of the particles produced by this process leads to longer residence times in the atmosphere. This is the case for sulfates that derive from SO_2 blasted into the stratosphere by large volcanic eruptions. It is also the case for acidic aerosols that contribute to acid deposition. Thus, SO_2 emitted from power plants in the United Kingdom can be deposited as sulfate far inland in continental Europe.

6.6 Sinks of aerosols

Aerosols can be removed from the atmosphere by dry and wet processes. Wet processes involve clouds and precipitation. The clean air and good visibility that often follow precipitation attests to its effectiveness in scavenging aerosols. We will discuss wet removal processes in Chapter 7.

Dry removal processes include coagulation of aerosols, sedimentation, and impaction onto surfaces (e.g., vegetation). We have already discussed coagulation, which is a sink for small aerosols but a source for larger aerosols. *Sedimentation* refers to the settling of aerosols due to the Earth's gravitational attraction. For particles in the form of spheres with diameter D, and if D is greater than the mean free path (λ) of the gas molecules in air ($\lambda \simeq 0.07\,\mu$m at 20°C and 1 atm), the *terminal settling velocity* v_s of a particle in still air is given by Stokes[5] equation

$$v_s = \frac{D^2 g}{18\mu}(\rho_p - \rho) \tag{6.13a}$$

where g is the acceleration due to gravity, μ the dynamic viscosity of air, ρ_p the density of the particle, and ρ the density of air. Since $\rho_p \gg \rho$

$$v_s \simeq \frac{D^2 g\, \rho_p}{18\mu} \tag{6.13b}$$

Substitution of numerical values into Eq. (6.13b) shows that sedimentation is significant only for aerosols with diameters greater that a few micrometers (Fig. 6.9). It is estimated that ~10% to 20% of the mass of aerosols removed from the atmosphere is by sedimentation.

Small aerosols, for which sedimentation over large distances is

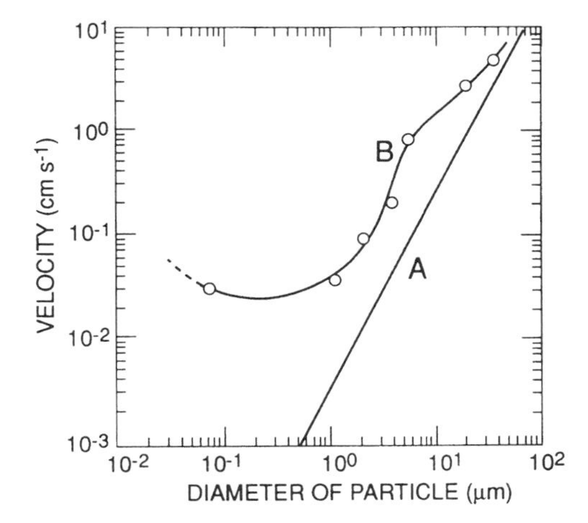

Figure 6.9. Aerosol terminal settling velocity (curve A) and deposition velocity onto a flat surface covered by grass (curve B). [Adapted from G. Hidy in *Chemistry of the Lower Atmosphere*, Ed. S. I. Rasool, Plenum Press, p. 155 (1973).]

ineffective, can impact onto obstacles (e.g., newly washed automobiles, vegetation etc.) if they are brought close to a surface by winds relative to the surface (e.g., the motion of the car) or by turbulent air motions. However, to make actual contact with the surface, the aerosols must be transported across a thin stagnant layer of air that borders the surface. This can be achieved by sedimentation (effective over short distances even for small particles), by the vertical motion of particles, or by diffusion. The diffusion may be random (*Brownian*) or directed (*phoretic*). Phoretic diffusion arises from ordered fluxes of molecules superimposed on disordered molecular motions. The two types of phoretic diffusion are important in the atmosphere: *diffusiophoresis*, which is produced by the flux of water molecules in the vapor phase during condensation or evaporation, and *thermophoresis*, which is produced by fluxes of heat during condensation or evaporation. For example, when water vapor condenses onto a surface (e.g., a leaf), diffusiophoresis tends to drive aerosols in the direction of the vapor flux (i.e., toward the surface). On the other hand, condensation releases heat that raises the temperature of the surface above the ambient air temperature. Therefore, thermophoretic forces tend to drive aerosols down the temperature gradient (i.e., away from a surface where condensation is occurring). The reverse holds for a surface from which water is evaporating. The magnitudes of the diffusiophoretic and thermophoretic forces depend on the size of the aerosols (see Exercise 1v in Appendix I).

As in the case of gases (see Section 5.5), we can define *deposition velocity* for aerosols onto a surface by

$$\frac{\text{deposition}}{\text{velocity}} = \frac{\text{flux of aerosols to the surface}}{\text{mass concentration of aerosols near the surface}} \quad (6.14)$$

Since the units of flux are mass per unit surface area per unit time (kg $m^{-2}s^{-1}$), and the units of mass concentration are mass per unit volume ($kg\,m^{-3}$), the right side of Eq. (6.14) has units of velocity ($m\,s^{-1}$).

Curve B in Figure 6.9 shows measurements of the deposition velocity of aerosols onto flat ground covered by grass. It can be seen that in this case the deposition velocity is always greater that the terminal settling velocity (curve A) and that the difference increases with decreasing aerosol size; this is due to turbulent diffusion enhancing aerosol deposition. The upturn in the deposition velocity with decreasing aerosol size below $D \simeq 0.3\,\mu m$ is due to Brownian motion.

6.7 Residence times of aerosols

As discussed in Section 2.2, the residence time of a material in the atmosphere depends on its sources and sinks. Shown in Figure 6.10(a) are estimates of the residence times of aerosols in the atmosphere as a function of their size. It can be seen that aerosols with diameter $<0.01\,\mu m$ have residence times $\lesssim 1$ day; the major removal mechanisms for particles of this size are phoretic diffusion to cloud particles and Brownian coagulation. Aerosols $\gtrsim 20\,\mu m$ diameter also have residence times $\lesssim 1$ day, but they are removed by sedimentation, impaction onto surfaces, and precipitation scavenging. On the other hand, large aerosols, with diameters between ~0.2 and $2\,\mu m$, have strong sources (coagulation of Aitken nuclei and particles left behind by the evaporation of cloud droplets) but weak sinks (indicated by the dashed lines in Fig. 6.10c). Consequently, large aerosols have relatively long residence times, reaching several hundred days in the upper troposphere, but precipitation scavenging and impaction reduces these residence times to a few tens of days in the middle and lower troposphere. It is for this reason that the so-called accumulation mode in aerosol surface area and volume plots occurs in the vicinity of the size range of large aerosols (e.g., see Figs. 6.5 and Fig. 6.10b).

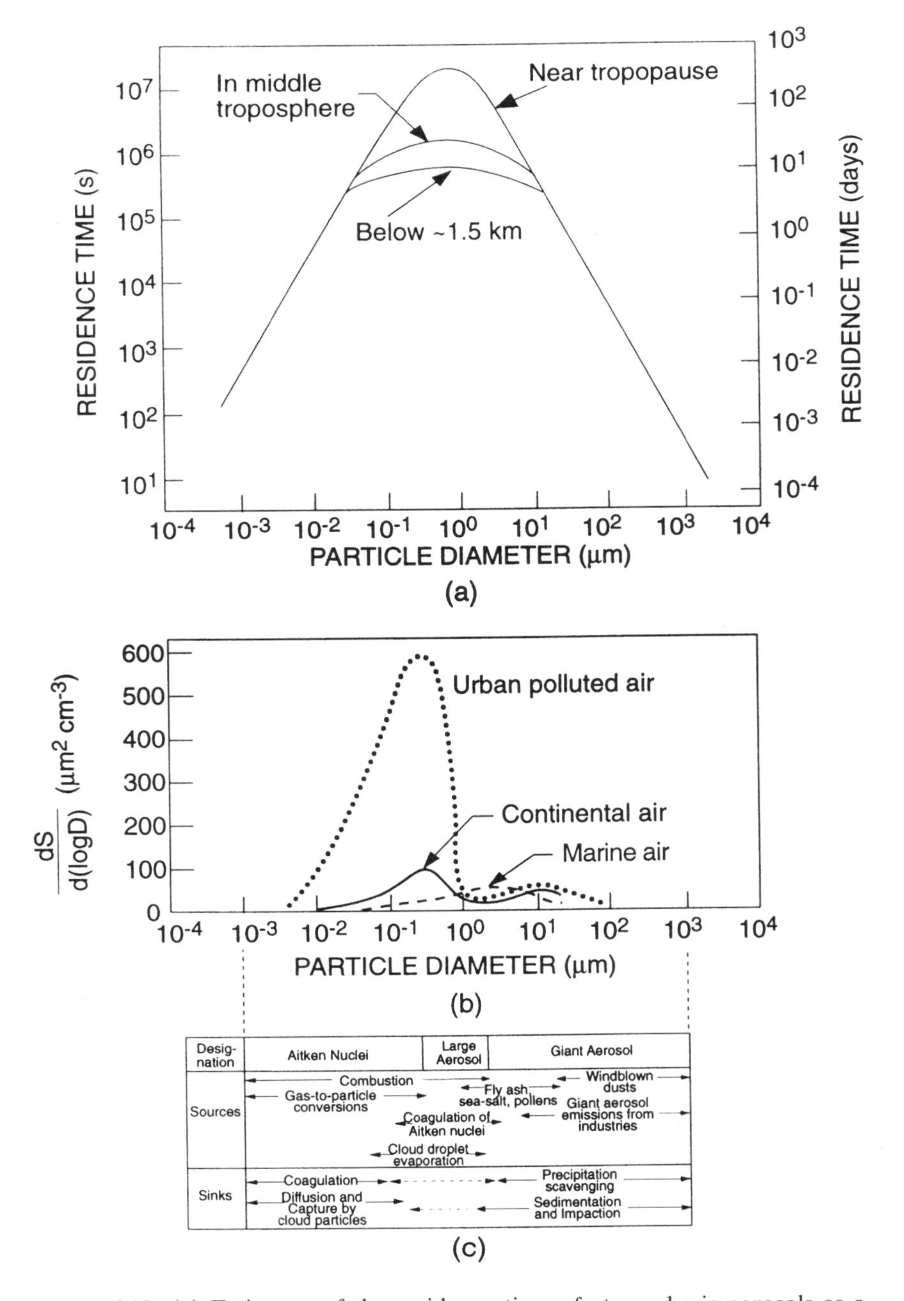

Figure 6.10. (a) Estimates of the residence time of atmospheric aerosols as a function of their diameters. [Adapted from R. Jaenicke, *Ber. Bunsen-Gesellschaft für Phys. Chemie*, **82**, 1198 (1978).] The relationship of the residence time to aerosol surface area distributions and the main sources and sinks of aerosols can be seen by comparing (a) with (b) and (c).

The residence times of aerosols in the middle and low troposphere are generally less than that of water vapor (~10 days). These short residence times produce large spatial and temporal variabilities in aerosol concentrations, especially near aerosol sources. The longer residence times of aerosols in the upper atmosphere, together with clouds serving as a fairly uniform source of aerosols, produces a more uniform and constant background of aerosols in this region (see Fig. 6.1).

6.8 Geographical distribution of aerosols

Currently there are insufficient measurements worldwide to provide detailed quantitative information on the global distribution of aerosols. However, the optical thickness of aerosols, which provides a measure of the effect of aerosols on depleting the amount of solar radiation reaching the Earth's surface (see Section 4.1), can be measured over the oceans (but not over land) from satellites. Average values of the aerosol optical thickness, derived in this way, for the four seasons are shown in Figure 6.11. Figure 6.11 shows that the most prominent areas of high values of the aerosol optical thickness are associated with continental sources (e.g., dust from the Sahara and the Middle East, and anthropogenic aerosols from Southeast Asia and the U.S. East Coast). However, there are also isolated patches of relatively high aerosol optical thicknesses that are consistent with more diffuse large-scale sources (e.g., from the oceans).

Using information on the sources, sinks, and residence times of aerosols, numerical models of global transport can be used to estimate the global distributions of aerosols. For example, Figure 6.12 shows the results of such computations for the distribution of soil dust aerosols. It can be seen from this figure that dust from the Sahara Desert can extend far into the Atlantic Ocean and even to the Americas; the rain forests of South America may be nourished by these dusts. During spring, dust from the Gobi Desert can be transported far over the Pacific Ocean.

6.9 Atmospheric effects of aerosols

Shown in Figure 6.13 are the approximate size ranges of aerosols that play a role in atmospheric electricity, air chemistry, atmospheric radiation, and clouds and precipitation processes. We have seen already

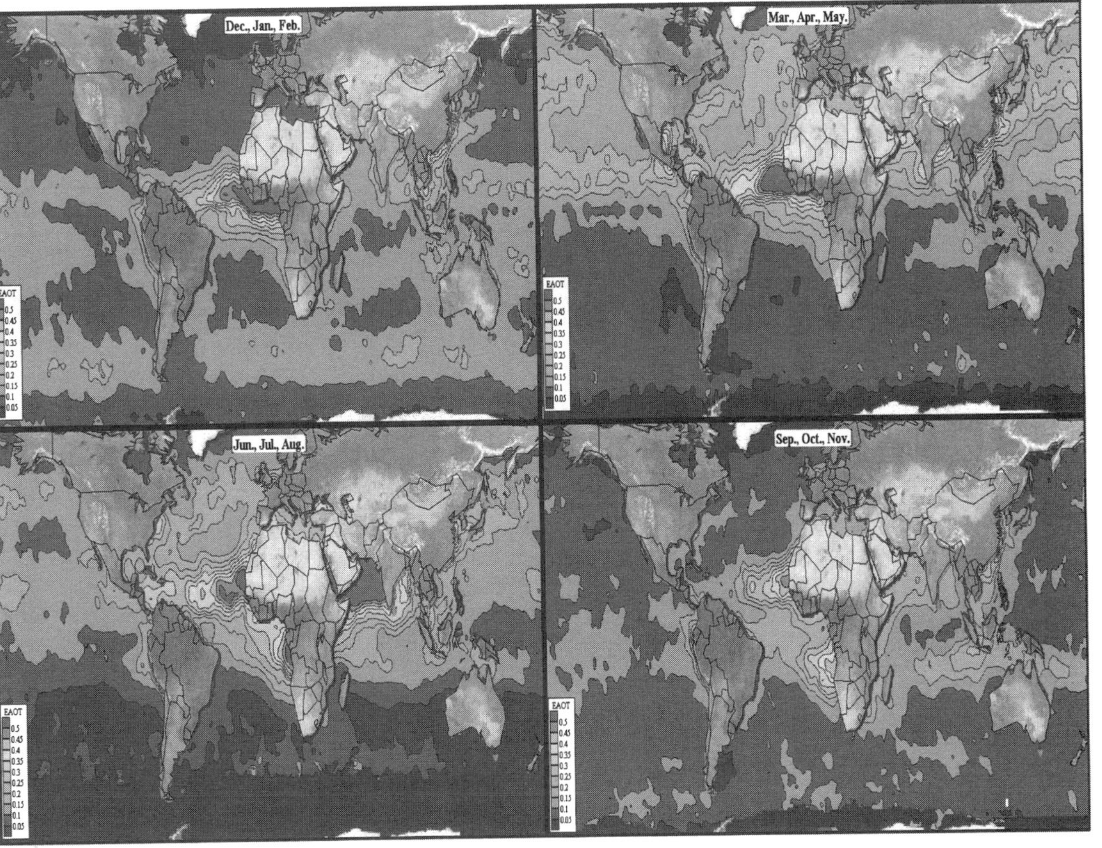

Figure 6.11. Radiatively equivalent aerosol optical thickness (EAOT × 1000) over the oceans derived from NOAA/AVHRR satellites for the four seasons. The figure incorporates data for the period July 1989–June 1991. Note that the continents are the major sources of aerosols over the ocean, but there is also evidence of oceanic contributions. [From R. Husar et al., *J. Geophys. Res.*, **102**, 16889 (1997). Copyright by the American Geophysical Union.] See color section found between page 118 and 119 for a color version of this figure.

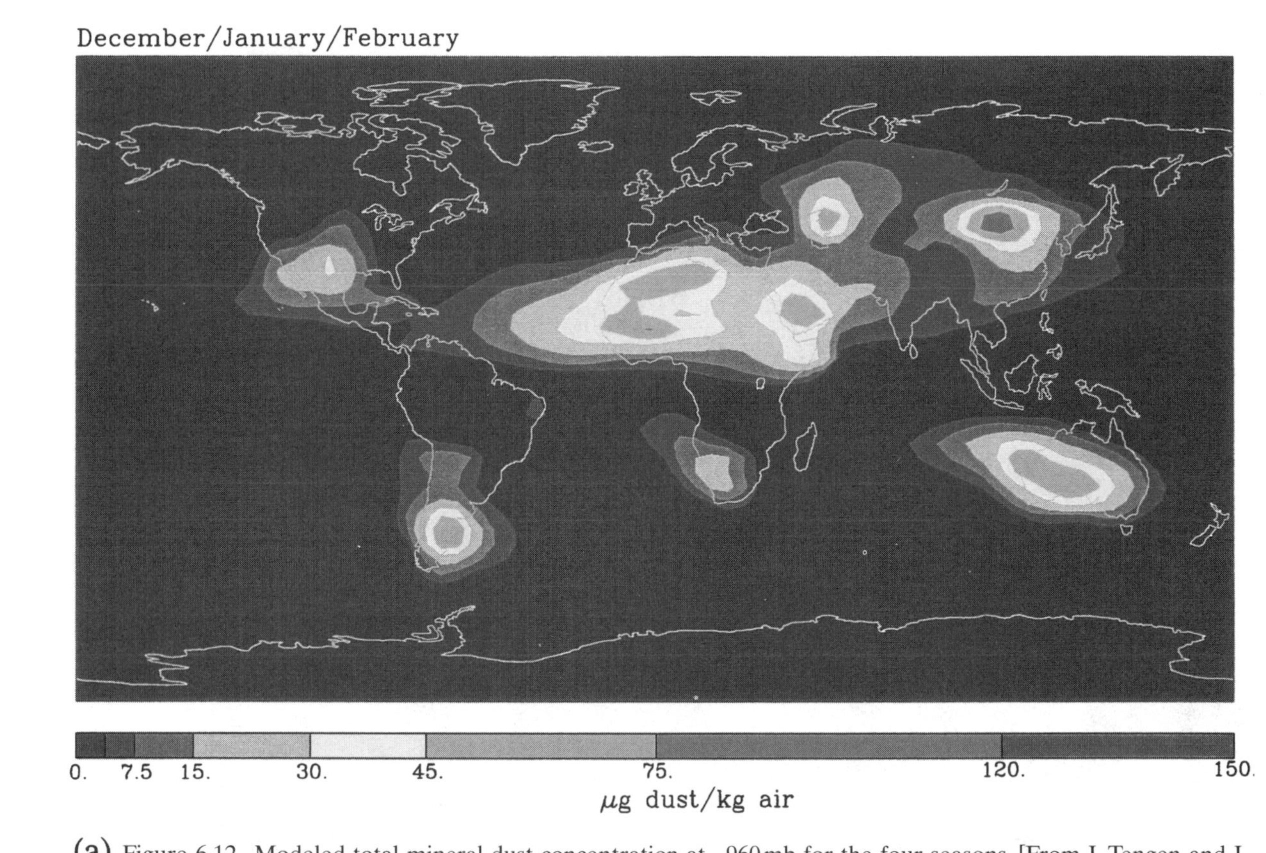

(a) Figure 6.12. Modeled total mineral dust concentration at ~960 mb for the four seasons. [From I. Tengen and I. Fung, *J. Geophys. Res.*, **99**, 22897 (1994). Copyright by the American Geophysical Union.] See color section found between page 118 and 119 for a color version of this figure.

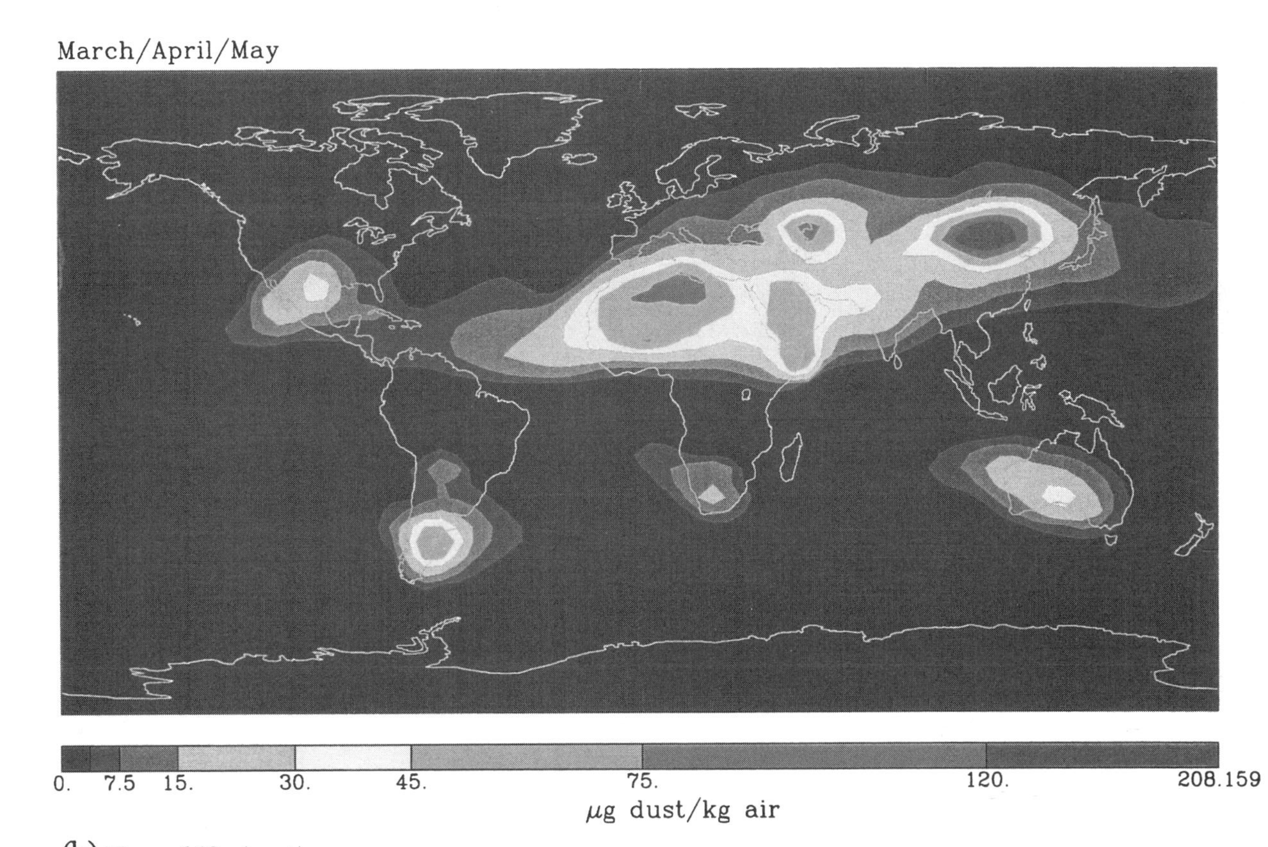

(b) Figure 6.12. *(cont.)*

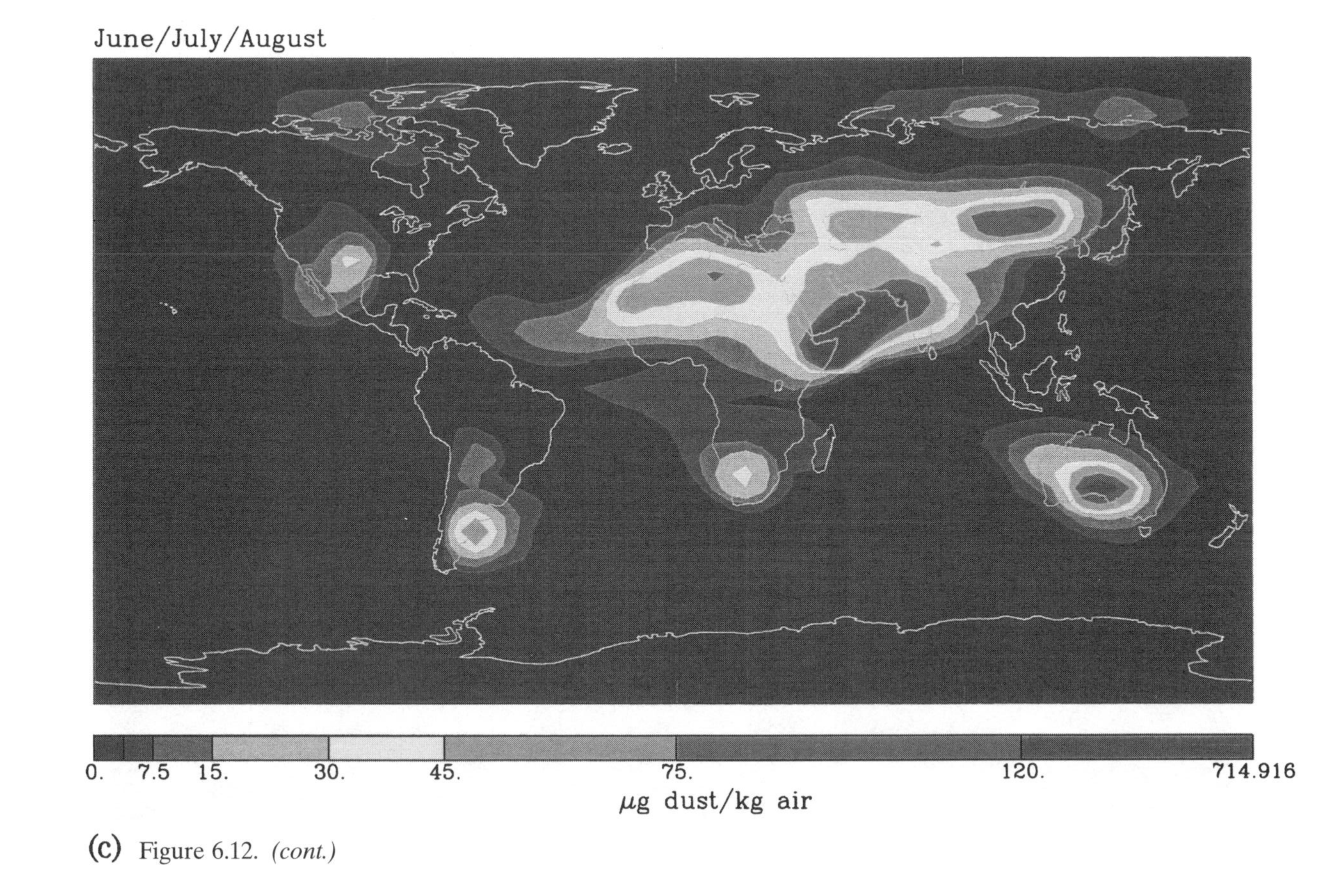

(c) Figure 6.12. *(cont.)*

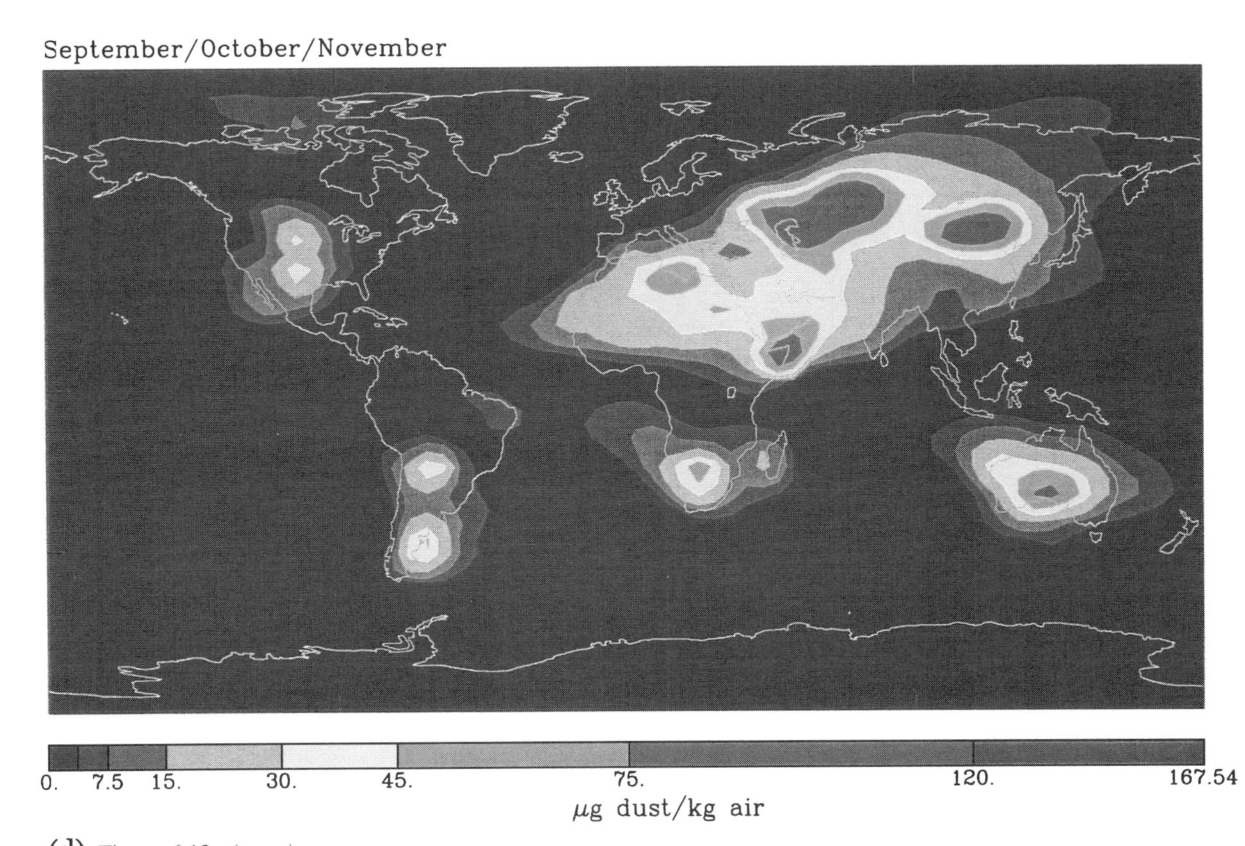

(d) Figure 6.12. *(cont.)*

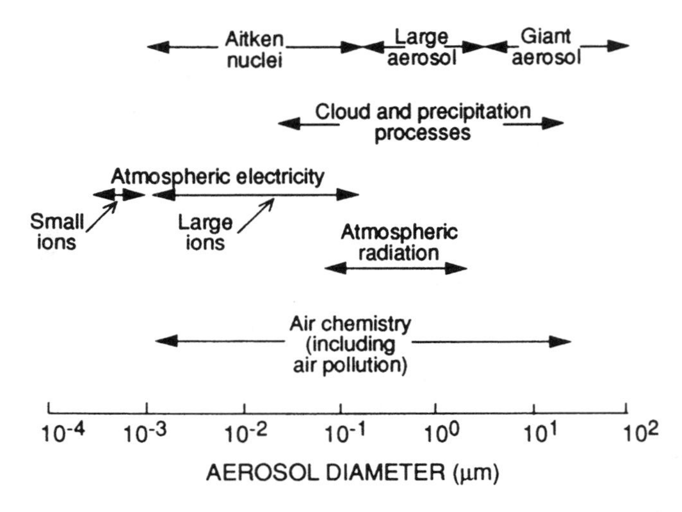

Figure 6.13. Approximate size ranges of aerosols of importance in various atmospheric phenomena.

that aerosols play an important role in the chemistry of the natural troposphere and in radiative transfer in the atmosphere. In the remaining chapters of this book we will see that they also play crucial roles in cloud and precipitation chemistry, air pollution, and stratospheric chemistry.

Exercises

See Exercises 1(t)–(v) and Exercises 27–32 in Appendix I.

Notes

1 John Aitken (1839–1919). Scottish physicist. In addition to pioneering work on atmospheric aerosols, he investigated cyclones, color, and color sensations.

2 For convenience, atmospheric aerosol particles are generally assumed to be spherical, although it is known that this is not always the case.

3 Christian E. Junge (1912–1996). German meteorologist. A pioneer in studies of atmospheric aerosols and trace gases.

4 Since, strictly speaking, we cannot take the logarithm of a dimensional quantity, such as D_i, we can think of $\log D_i$ in Eqs. (6.6) and (6.7) as $\log(D_i/1)$, where the "1" refers to a reference particle with diameter 1 μm. Note also that because σ_g is a ratio of diameters (see Eq. 6.6b) it is a pure number, unlike the regular standard deviation.

5 Sir George Stokes (1819–1903). Mathematician and physicist, born in Ireland. Close friend of Lord Kelvin. Developed modern theories for the motion of viscous fluids and for diffusion. Discovered fluorescence (1852). One of the founders of the science of geodesy.

7
Cloud and precipitation chemistry

Early studies of atmospheric chemistry emphasized trace gases and homogeneous gas-phase reactions. However, in the latter part of the twentieth century, increasing attention was given to atmospheric aerosols (see Chapter 6), chemical reactions on aerosol surfaces (see Chapter 10), and the role of clouds in atmospheric chemistry. In this chapter we consider how clouds remove particles and gases from the air, some of the chemical reactions that can occur within cloud droplets, and how these processes modify the chemical composition of cloud water and precipitation, as well as some other important properties of clouds. Finally, we will discuss the effects of cloud processing on modifying atmospheric aerosols.

7.1 Overview

We will organize our discussion of cloud and precipitation chemistry around the processes illustrated schematically in Figure 7.1.

Clouds form when air becomes slightly supersaturated with respect to liquid water (or in some cases with respect to ice). The most common means by which this is achieved in the atmosphere is through the ascent of air parcels, which results in the expansion and cooling of the air below its dew point. When the air becomes slightly supersaturated with respect to water (by a few tenths of 1%), water vapor begins to condense onto some of the particles in the air to form a cloud of small water droplets.[1] This process, by which a portion of the atmospheric aerosol (called *cloud condensation nuclei*, or CCN) nucleate the formation of cloud droplets, and are thereby incorporated into cloud water as insoluble and soluble components, is referred to as *nucleation scavenging*.

Particles that are ingested into a cloud but do not serve as CCN and

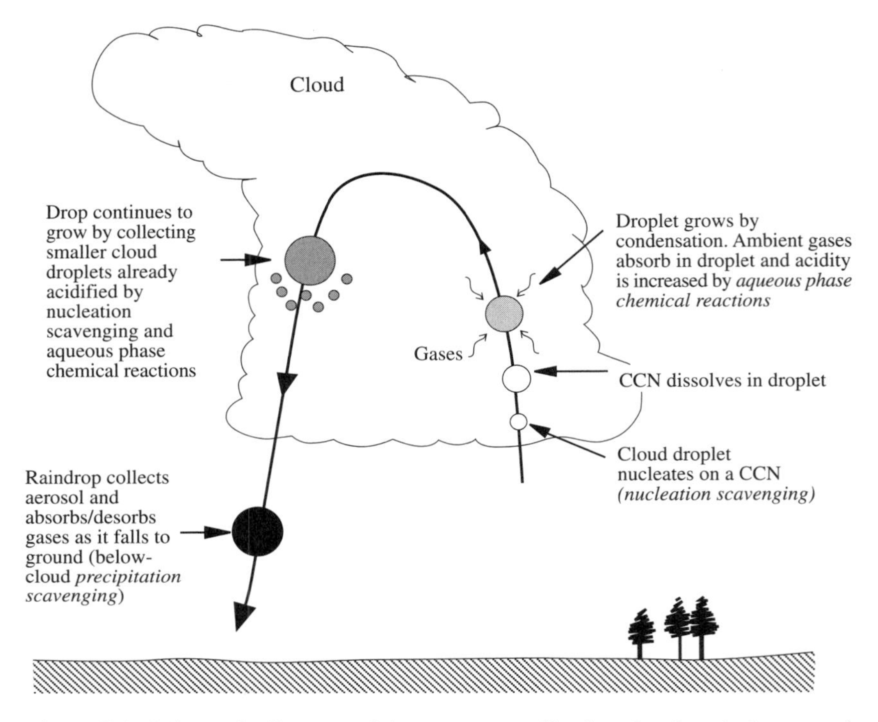

Figure 7.1. Schematic diagram of the processes affecting the chemical compositions of cloud droplets and rain. Not drawn to scale.

therefore reside in the air between cloud droplets, are called *cloud interstitial aerosol* (CIA). The CIA diminish very slowly in number as a cloud ages due to their coagulation with cloud droplets.

As cloud droplets are carried up into a cloud, they grow to tens of micrometers in radius by the condensation of water in the water-supersaturated air. At the same time, various trace gases in the air, particularly those with high solubilities, dissolve in the cloud droplets. *Aqueous-phase chemical reactions* may then occur in the droplets at much faster rates than they occur in the gas phase.

If cloud droplets attain radii in excess of about 20 μm, they can begin to grow rapidly by colliding and coalescing with smaller droplets. This collision–coalescence process can lead to the formation of rain droplets. Raindrops have chemical compositions that are determined by the compositions of the droplets from which they are formed. As a raindrop falls through a cloud and below cloud base, it will collect some of the aerosol lying in its path and may absorb and desorb various gases. These processes are referred to as *precipitation scavenging*.

In the next several sections we will consider in more detail these four processes and depicted in Figure 7.1 (namely, *nucleation scavenging*, *dissolution of gases*, and *aqueous-phase chemical reactions in clouds*, as well as *precipitation scavenging*), which together determine the chemical compositions of cloud water and raindrops.

7.2 Cloud condensation nuclei and nucleation scavenging

As we have seen in Chapter 6, the atmosphere contains many particles that range in size from submicrometer to tens of micrometers. *Wettable*[2] particles of sufficient size can serve as CCN. This can be seen from *Kelvin's*[3] *equation*, which gives the equilibrium vapor pressure just above the surface of a water droplet of radius r as[4]

$$e = e_s \exp\left(\frac{2\sigma}{nkTr}\right) \tag{7.1}$$

where e_s is the saturation vapor pressure over a plane surface of pure water, σ the interfacial or surface energy of water (defined as the work required to create unit area of interface between water and air), n the number of molecules per unit volume of water, k Boltzmann's constant (1.381×10^{-23} J deg^{-1} molecule^{-1}), and T the temperature in degrees Kelvin. The water supersaturation (a percentage) corresponding to e is defined as $[(e/e_s) - 1]100$. This is shown as the dashed curve in Figure 7.2, where it can be seen that the smaller the droplet the greater the water supersaturation just above its surface. For a droplet to survive or grow, the supersaturation in the ambient air would have to be at least as great as that just above its surface.

If sufficient numbers of water molecules condense onto a completely wettable but water-insoluble particle, say, 0.5 μm in radius, to form a thin film of water over the surface of the particle, the supersaturation just above the water surface will be given by Kelvin's equation, namely, about 0.4% (point X in Fig. 7.2). Hence, if the supersaturation of the ambient air is greater than 0.4%, water will continue to condense onto the particle and a cloud droplet several micrometers in size will form. It can be seen from the dashed curve in Figure 7.2 that the larger the size of the original particle, the lower will be the supersaturation of the ambient air required for a droplet to form on the particle.

A different situation occurs if the particle onto which water condenses

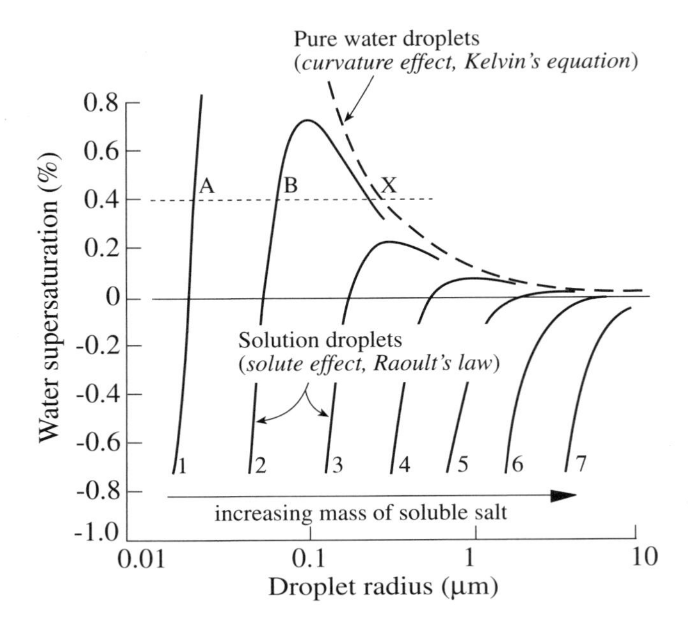

Figure 7.2. Equilibrium supersaturation (with respect to a plane surface of water) over droplets of pure water (dashed curve), and over droplets containing dissolved material, as a function of droplet radius. Curves labeled 1–7 represent increasingly greater amounts of dissolved material in the droplet; they are called Köhler curves.

is partially water soluble, so that some of it dissolves to form a small solution droplet. The equilibrium vapor pressure over a solution droplet is less than that over a pure water droplet of the same size for the following reason. The equilibrium vapor pressure is proportional to the concentration of water molecules on the surface of the droplet. In a solution droplet some of the surface molecular sites are occupied by the molecules of salt (or ions, if the salt dissociates); therefore, the vapor pressure is reduced by the presence of the solute. The fractional reduction in vapor pressure is given by the relation[5]

$$\frac{e'}{e} = f \tag{7.2}$$

where e' is the saturation vapor pressure over a solution droplet containing a mole fraction f of pure water, and e is the saturation vapor pressure over a pure water droplet of the same size and at the same temperature. The *mole fraction of pure water* is defined as the number of moles of pure water in the solution divided by the total number of moles (pure water plus solute) in the solution.

Let us consider now a solution droplet of radius r that contains a mass m of a dissolved salt of molecular weight M_s. If each molecule of the salt dissociates in water into i ions, the effective number of moles of the salt in the droplet is im/M_s. If the density of the solution is ρ', and the molecular weight of water M_w, the number of moles of pure water in the droplet is $(\frac{4}{3}\pi r^3 \rho' - m)/M_w$. Therefore, the mole fraction of water is

$$f = \frac{\dfrac{\left(\frac{4}{3}\pi r^3 \rho' - m\right)}{M_w}}{\dfrac{\left(\frac{4}{3}\pi r^3 \rho' - m\right)}{M_w} + im/M_s} = \left[1 + \frac{imM_w}{M_s\left(\frac{4}{3}\pi r^3 \rho' - m\right)}\right]^{-1} \tag{7.3}$$

The vapor pressure e over a pure water droplet of radius r is given by Eq. (7.1). Combining Eqs. (7.1)–(7.3), the following expression is obtained for the saturation vapor pressure e' over a solution droplet of radius r

$$\frac{e'}{e_s} = \left[\exp\frac{2\sigma'}{n'kTr}\right]\left[1 + \frac{imM_w}{M_s\left(\frac{4}{3}\pi r^3 \rho' - m\right)}\right]^{-1} \tag{7.4}$$

where the dashed superscript indicates values appropriate to the solution. Equation (7.4) may be used to calculate the saturation vapor pressure e' [or relative humidity $100\, e'/e_s$, or supersaturation $(e'/e_s - 1)100$] in the air just above the surface of a solution droplet.

If we use Eq. (7.4) to plot the variation of the relative humidity (or supersaturation) of the air just above a solution droplet as a function of its radius r, we obtain what is referred to as a *Köhler*[6] *curve*. Several such curves are shown by the solid lines in Figure 7.2. If a solution droplet is below a certain size, the saturated vapor pressure above the droplet will be less than that above a plane surface of pure water at the same temperature. If a droplet containing a fixed mass of solute increases in size, the strength of the solution will decrease. Eventually the Kelvin curvature effect (which increases the saturation vapor pressure above the droplet) dominates over the solute effect (which decreases the saturation vapor pressure). When the droplet becomes large enough, the saturation vapor pressure above its surface will become essentially the same as that over a pure water droplet with the same radius and at the same temperature.

The Köhler curves can be used to determine how a solution droplet will grow in an environment at a fixed supersaturation. For example, we have seen that in an environment with a supersaturation of 0.4%, a water-insoluble particle with a radius less than about 0.5 μm cannot serve as a CCN to form a cloud droplet. This is because if a thin layer of water were on its surface the supersaturation above this layer would be >0.4%, so the condensed water would evaporate. However, in this same environment, with a supersaturation of 0.4%, droplets containing dissolved material (of dry radius much less than 0.5 μm) represented by curve 1 in Figure 7.2, would increase in size up to point A, at which point the droplet would be in equilibrium with the ambient vapor pressure. Similarly, a solution droplet represented by curve 2 would grow to point B. In the atmosphere, droplets that are in equilibrium states such as A and B (to the left of the peaks in their Köhler curve) are referred to as *unactivated drops* or *haze.* Haze can significantly decrease the intensity of solar radiation reaching the earth; it can also cause significant decreases in visibility. Note that water-soluble particles can form haze at relative humidities well below 100% (corresponding to negative values of supersaturation in Fig. 7.2).

Next consider a solution droplet represented by curve 3 in Figure 7.2, which is exposed to an ambient supersaturation of 0.4%. In this case, the peak in the Köhler curve lies below 0.4% supersaturation. Therefore, the droplet can grow by condensation up the left-hand side of the Köhler curve, over the peak in this curve, and down the right-hand side of the curve. The droplet is now said to be *activated*, because it has formed a cloud droplet, initially several micrometers in radius, which can grow to larger sizes.

In general, a water-soluble particle will be activated in ambient air with supersaturation S if $S > S_c$, where S_c is the peak value of the supersaturation given by the Köhler curve for the particle. S_c is given by

$$S_c = \left(\frac{2.5 \times 10^5}{\text{number of soluble ions}} \right)^{1/2} \tag{7.5}$$

Exercise 7.1. What are the values of the saturation relative humidity and supersaturation just above a pure water droplet of radius 0.05 μm at 10°C? Assume that the values of σ and n in Eq. (7.1) are 0.074 N m^{-1} and 34.3 × 10^{28} m^{-3}, respectively.

Solution. From Eq. (7.1)

$$e = e_s \exp\left(\frac{2 \times 0.074}{3.3 \times 10^{28} \times 1.381 \times 10^{-23} \times 283 \times 0.05 \times 10^{-6}}\right)$$

$$\frac{e}{e_s} = \exp(0.023)$$

$$\frac{e}{e_s} = 1.02$$

Therefore, the saturation relative humidity (RH) over the droplet is

$$RH \equiv \frac{e}{e_s} 100 = 102\%$$

and the supersaturation (SS) is

$$SS \equiv \left(\frac{e}{e_s} - 1\right) 100 = 2\%$$

Exercise 7.2. If 10^{-19} kg of sodium chloride dissolves in the 0.05 μm radius droplet in Exercise 7.1, what is the saturation relative humidity just above the solution droplet? Assume that the density of the solution is 10^3 kg m^{-3}, that the values of σ' and n' are the same as those for pure water given in Exercise 7.1, and T is 283°K.

Solution. From Eq. (7.4)

$$\frac{e'}{e_s} = \exp\left(\frac{2\sigma'}{n'kTr}\right)\left[1 + \frac{imM_w}{M_s\left(\frac{4}{3}\pi r^3 \rho' - m\right)}\right]^{-1}$$

where

$$\exp\left(\frac{2\sigma'}{n'kTr}\right) = 1.02$$

(see Exercise 7.1), and $i = 2$ for NaCl, $m = 10^{-19}$ kg, $M_w = 18$, $M_{s=\text{NaCl}} =$ 22.99 + 35.45 = 58.44, $\rho' = 10^3$ kg m^{-3}, and $r = 0.05 \times 10^{-6}$ m. Therefore,

$$\frac{e'}{e_s} = 1.02\left[1 + \frac{2\times10^{-19}\times18}{58.44\left[\frac{4}{3}\pi(0.05\times10^{-6})^3 10^3 - 10^{-19}\right]}\right]^{-1}$$

$$= 1.02[1.19]^{-1}$$

$$= 0.86$$

Therefore, the saturation relative humidity (RH) above the droplet is

$$RH = 100\frac{e'}{e_s} = 86\%$$

It follows from the preceding discussion that the larger the size of a particle, and the larger its water solubility (or, if it is water insoluble, the more readily it is wetted by water), the lower will be the supersaturation at which it can serve as a CCN. To act as CCN at 1% supersaturation, completely wettable but water insoluble particles need to be at least about 0.1 μm in radius, whereas water-soluble particles can be as small as about 0.01 μm in radius. Because of these restrictions, only a small fraction of atmospheric aerosol serve as CCN (about 1% in continental air and 10% to 20% in marine air). Most CCN probably consist of a mixture of water-soluble and water-insoluble components (so-called *mixed nuclei*).

Worldwide measurements of CCN concentrations have not revealed any systematic latitudinal or seasonal variations. However, near the Earth's surface, continental air masses are generally significantly richer in CCN than are marine air masses. For example, at 1% supersaturation the concentration of CCN in continental air is typically on the order of 500 cm^{-3}, while in marine air it is about 100 cm^{-3}. Concentrations of CCN over land decline by about a factor of five between the surface and 5 km; over the same height interval concentrations of CCN over the ocean remain fairly constant.

The observations described provide some clues as to the origins of CCN. It appears that the land acts as one source of CCN because of the higher concentrations of CCN over the land and their decrease in concentration with altitude. Some of the soil particles and dusts that enter the atmosphere probably serve as CCN, but they do not appear to be a dominant source. Forest fires are sources of CCN; it has been estimated that the rate of production of CCN from burning vegetable matter is on

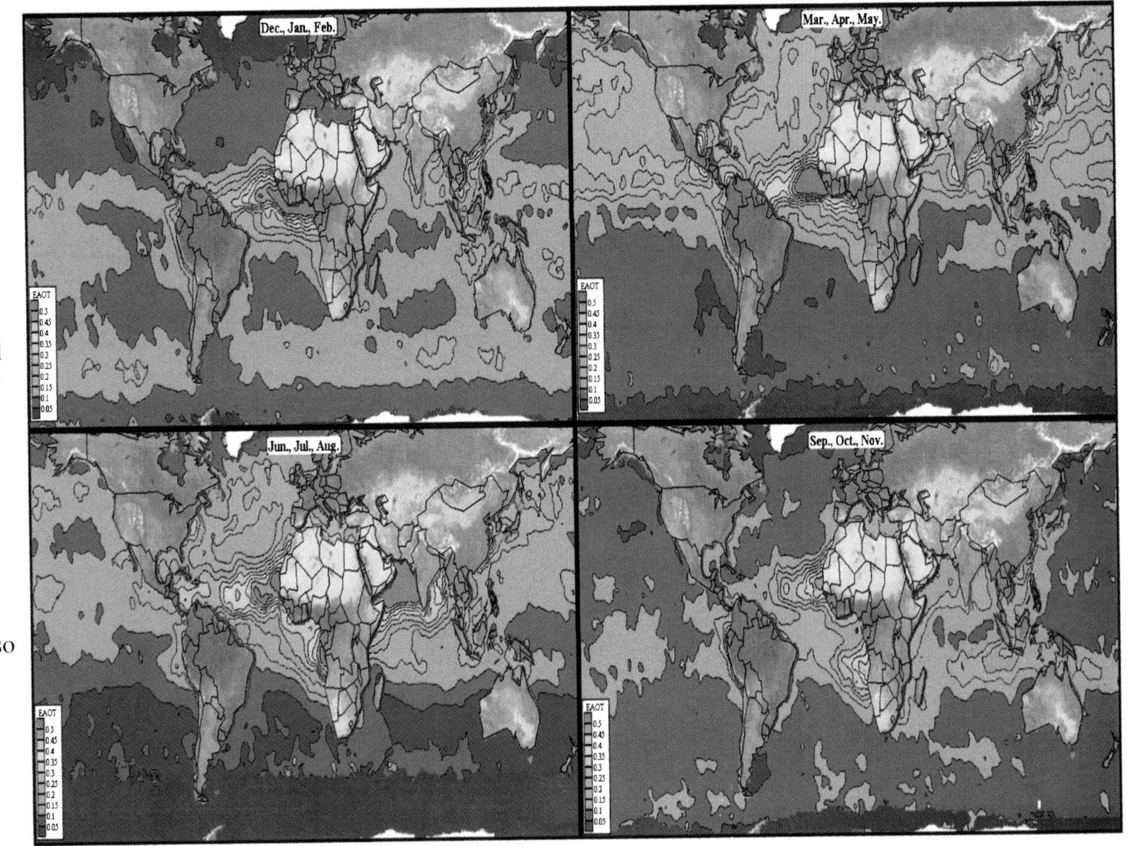

Figure 6.11. Radiatively equivalent aerosol optical thickness (EAOT × 1000) over the oceans derived from NOAA/AVHRR satellites for the four seasons. The figure incorporates data for the period July 1989–June 1991. Note that the continents are the major sources of aerosols over the ocean, but there is also evidence of oceanic contributions. [From R. Husar et al., *J. Geophys. Res.*, **102**, 16889 (1997). Copyright by the American Geophysical Union.]

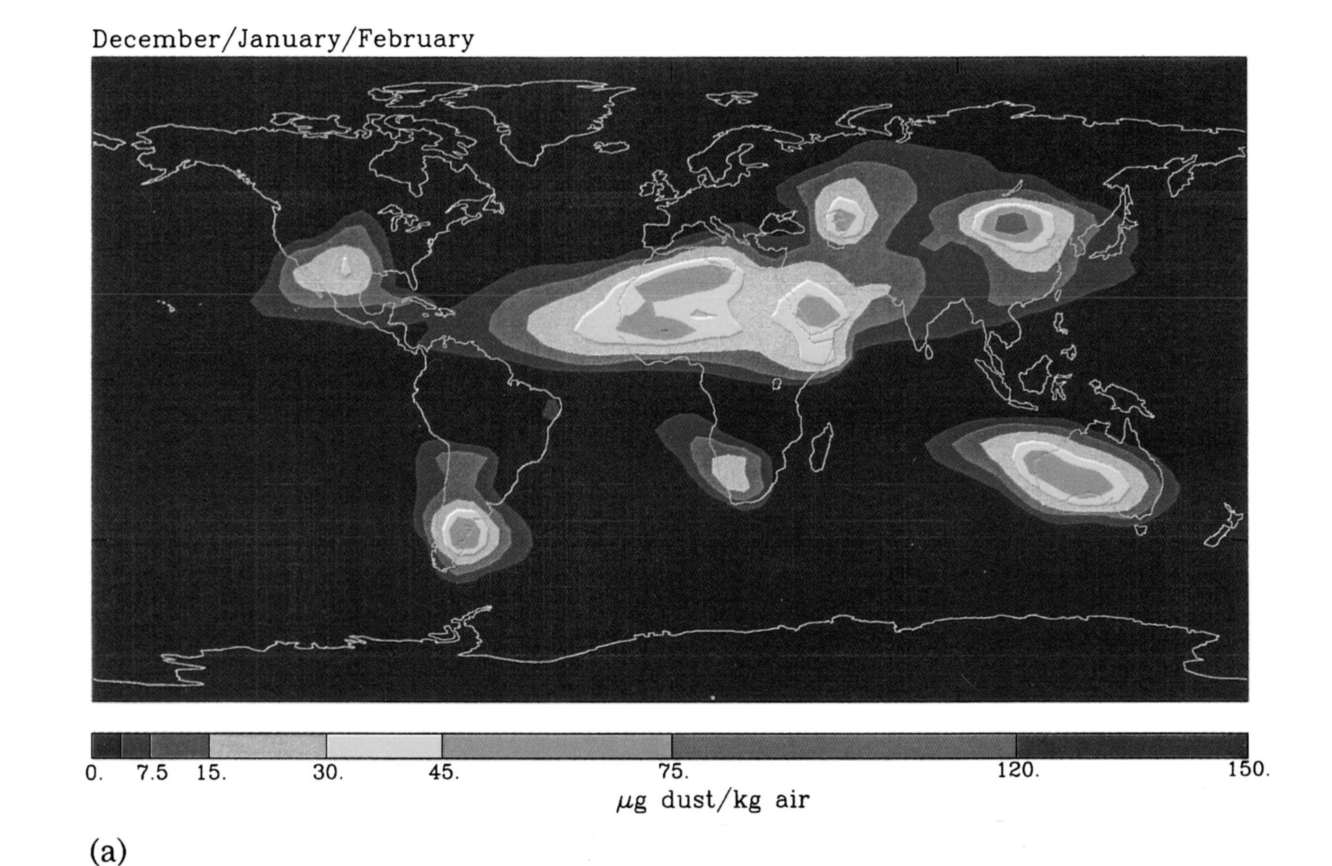

Figure 6.12. Modeled total mineral dust concentration at ~960 mb for the four seasons. [From I. Tengen and I. Fung, *J. Geophys. Res.*, **99**, 22897 (1994). Copyright by the American Geophysical Union.]

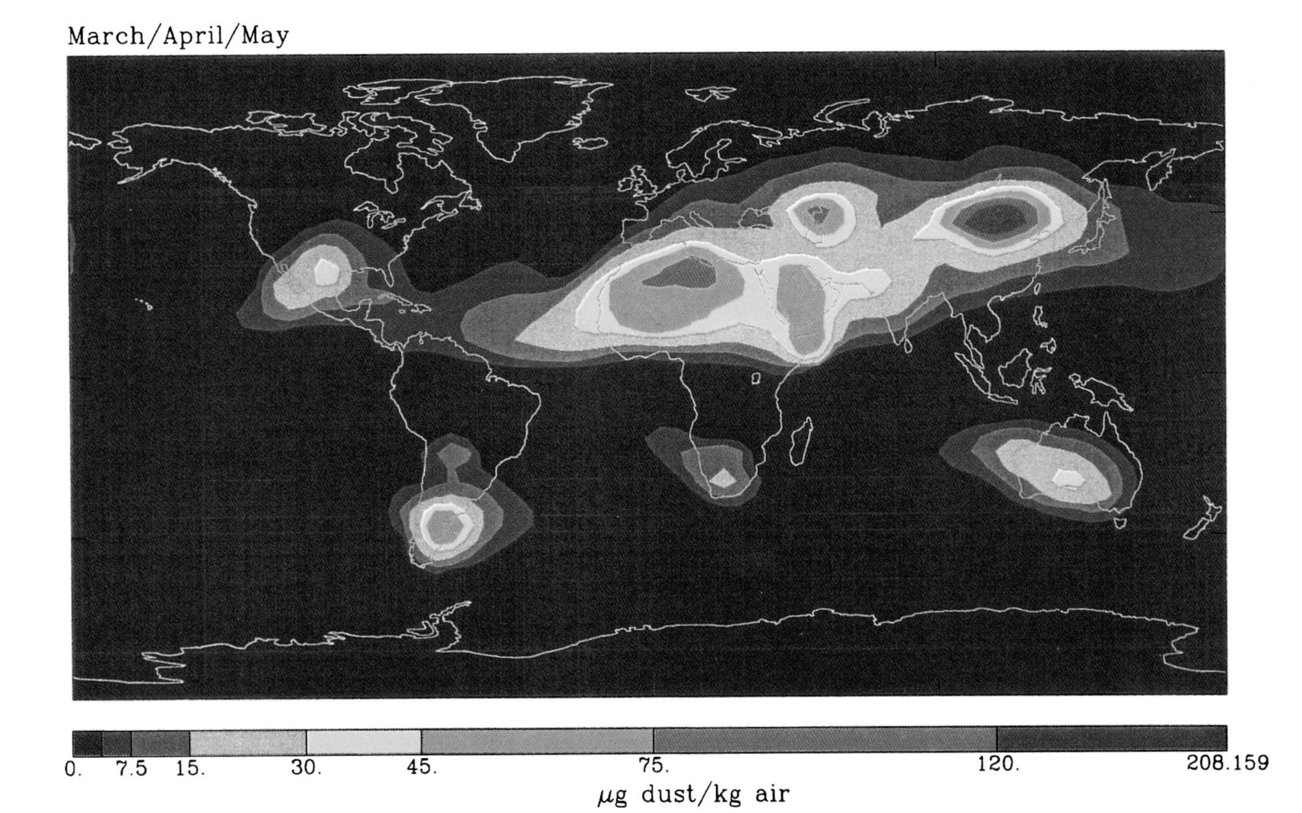

Figure 6.12. *(cont.)*

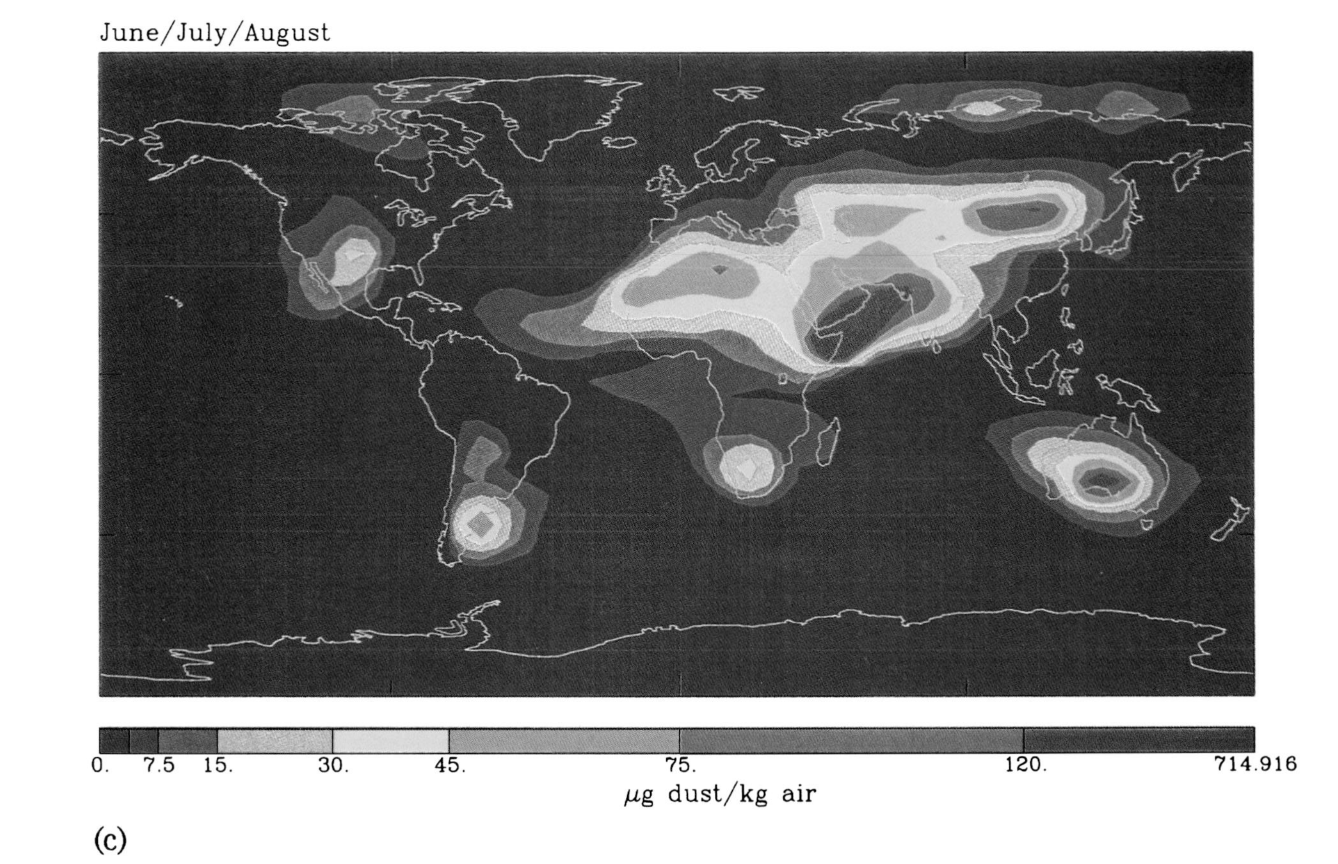

Figure 6.12. *(cont.)*

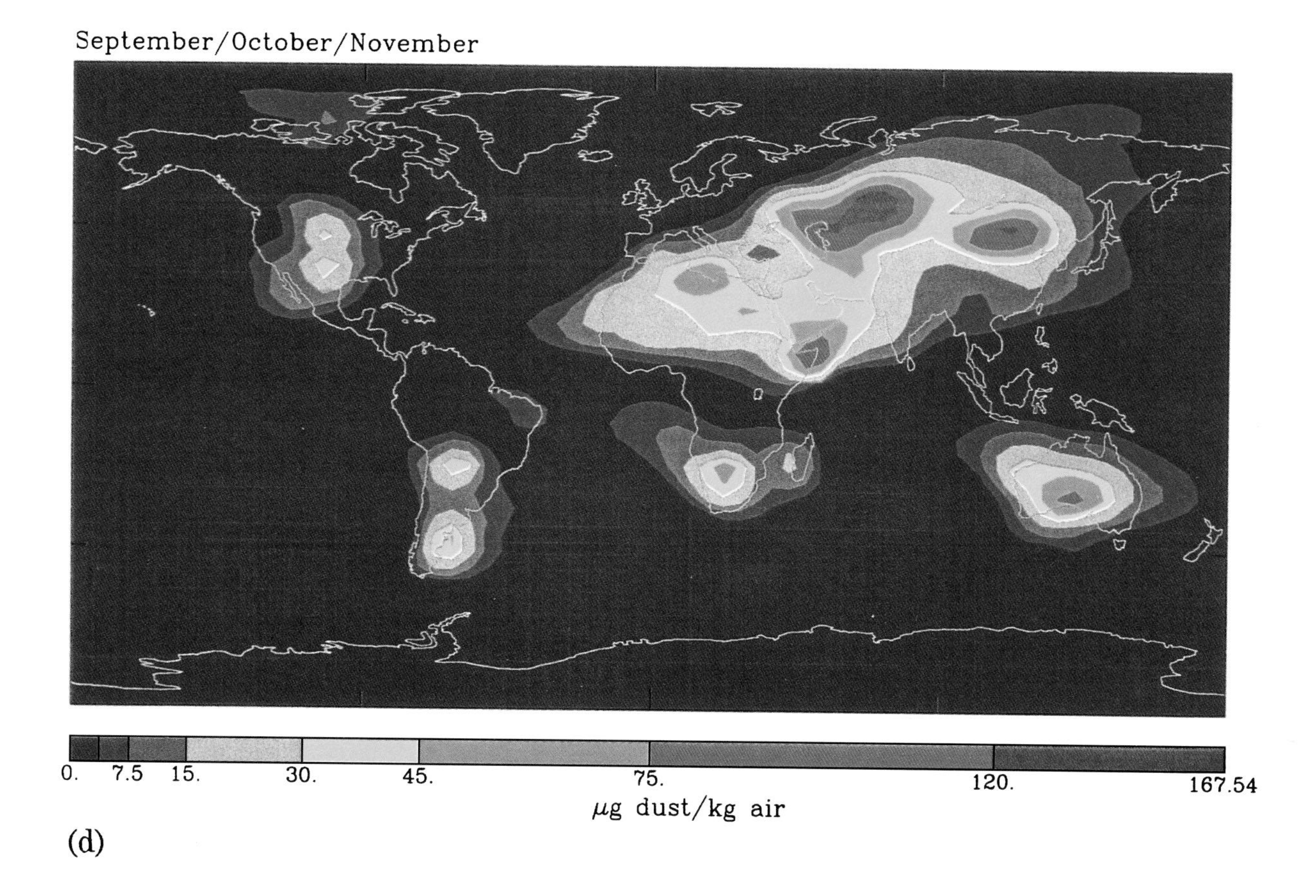

Figure 6.12. *(cont.)*

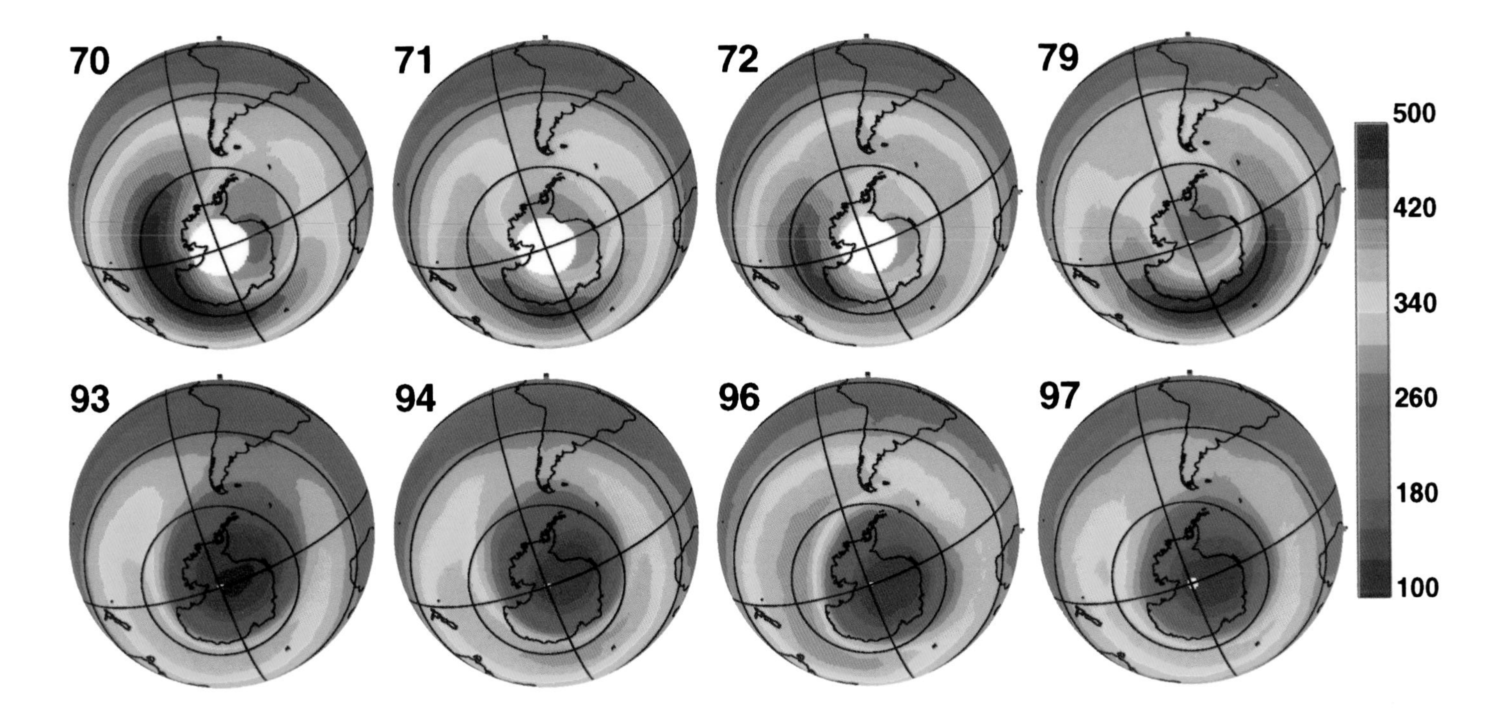

Figure 10.3. Satellite observations of the Antarctic ozone hole in the southern hemisphere during October for the years 1970–1997. The color scale is in Dobson units. (Courtesy P. Newman, NASA Goddard Space Flight Center.)

the order of 10^{12} to 10^{15} per kilogram of material consumed. Certain industries are sources of CCN (e.g., paper mills), but not all industrial pollutants act as CCN.

There appears to be a widespread and probably a fairly uniform source of CCN over both the oceans and the land. An important component of these CCN appears to be non–sea-salt sulfate. Over the oceans this sulfate may originate as follows. The gaseous sulfur species dimethylsulfide (DMS), which is produced by phytoplankton in the ocean, can be oxidized in the air to SO_2 and methane sulphonate. These gases can then be oxidized in the air to form sulfate particles. However, because of their small sizes, these particles can generally serve as CCN only at the relatively high supersaturations attained in convective clouds. More sulfate may be produced within cloud droplets by various aqueous-phase chemical reactions (see Section 7.4). Therefore, when a convective cloud evaporates, it will leave behind sulfate particles that are larger than those that served as the CCN on which the original cloud droplets formed. Consequently, these particles can serve as CCN at the lower supersaturations associated with stratiform clouds. Sea-salt particles and organic compounds, ejected into the air by air bubbles bursting at the ocean surface (see Fig. 6.7), may make significant contributions to CCN in marine boundary-layer air. These various processes are illustrated schematically in Figure 7.3.

Over land, in the northern hemisphere at least, the sulfur gas budget is dominated by SO_2 emissions from anthropogenic sources, which can be oxidized to form sulfate aerosols. This, together with other sources of potential CCN over land, accounts for the much higher CCN concentrations over the continents than over the remote oceans. Since the liquid water contents of continental and marine clouds of similar type (e.g., cumulus or stratus) do not differ appreciably, but the former have higher concentrations of droplets than the latter, the average size of the droplets in a continental cloud will generally be less than in a marine cloud of the same type. This dichotomy gives rise to some important consequences[7]:

- Because of the larger average sizes of droplets in clouds over the ocean, these clouds are more likely to precipitate than similar clouds over land.
- The reflection of solar radiation will generally be greater for clouds over land than for clouds of similar type over the ocean, because of the greater surface area of the droplets.

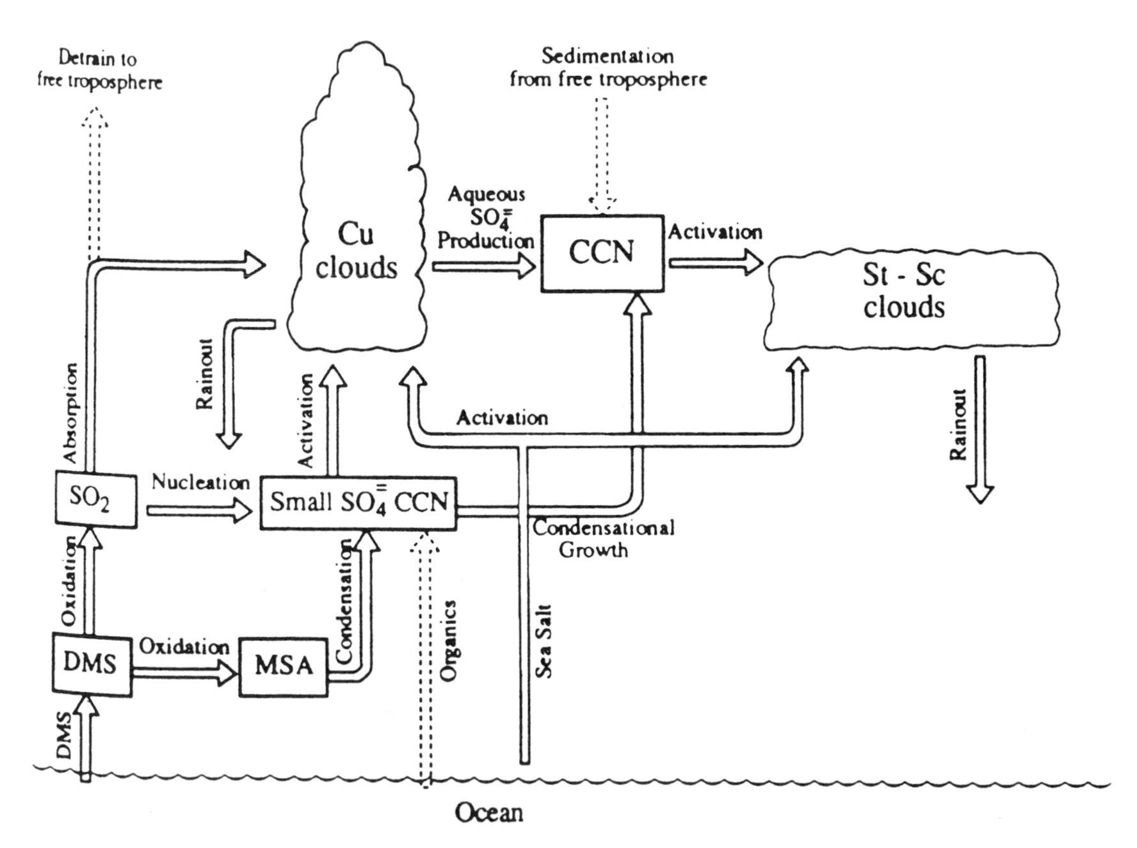

Figure 7.3. Schematic diagram of the sources and sinks of CCN in relatively clean oceanic air.

- The concentrations and sizes of droplets in clouds over the ocean are more susceptible to modification by perturbations in CCN than are similar cloud types over land.

These considerations have led to the hypothesis that changes in sea-surface temperature (e.g., produced by global warming) might possibly change DMS emissions from the oceans, which in turn could lead to changes in the amount of solar radiation scattered back into space by marine clouds. Of particular importance from this point of view are marine stratiform clouds, because these cover ~25% of the world's oceans and play an important role in the Earth's radiation balance.[8]

7.3 Dissolution of gases in cloud droplets[9]

As soon as water begins to condense to form a cloud (or haze or fog), gases in the surrounding air will begin to dissolve in the droplets. If the water is in equilibrium with a gas, the amount of solute (in moles) present in a given amount (say, 1 liter) of the saturated solution, called the *solubility* (C_g) of the gas, is given by *Henry's law*[10]:

$$C_g = k_H \, p_g$$

where, p_g is the partial pressure (in atmospheres) of the gas, and k_H is a temperature-dependent proportionally constant called the *Henry's law constant* or *Henry's law coefficient* (units: mole liter^{-1} atm^{-1}). Values of k_H for some common atmospheric gases are given in Figure 7.4.

The value of k_H generally increases with decreasing temperature. The temperature dependence of k_H is given approximately by

$$k_H(T_2) = k_H(T_1)\exp\left[\frac{\Delta \overline{H}_{rx}}{R^*}\left(\frac{1}{T_1} - \frac{1}{T_2}\right)\right] \tag{7.7}$$

where T_1 and T_2 are temperatures (in K), R^* the universal gas constant, and $\Delta \overline{H}_{rx}$ the molar enthalpy (or heat) of reaction at constant temperature and pressure.

Most of the values of k_H given in Figure 7.4 are based on measurements at partial vapor pressures far above atmospheric values. Therefore, considerable extrapolations are involved when they are applied to atmospheric conditions. Indeed, some of the very large values of k_H shown in Figure 7.4 cannot be applied to conditions in the atmosphere. On the other hand, k_H accounts for only the physical solubility of a gas in its undissociated form. Therefore, if a dissolved gas is involved in

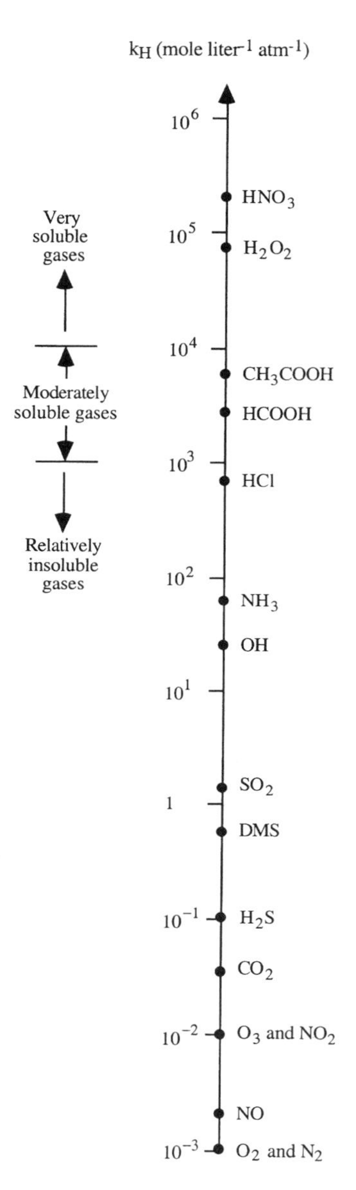

Figure 7.4. Values of the Henry's law constant (k_H) for some atmospheric gases in water at 25°C. These values account for the physical solubility of the gas only; any hydrolysis or other chemical reactions in the aqueous phase are not accounted for by these values. (The temperature dependence of k_H is discussed in Section 4.3 of Hobbs (2000). See Appendix VI.)

chemical reactions in the droplet (see Section 7.4), this will lead to a solubility that is greater than that indicated by the value of k_H for the gas. A quantitative approach to this problem is illustrated by the following exercise.

Exercise 7.3. By considering the dissolution of carbon dioxide gas in water with a Henry's law constant k_H (CO_2), and the subsequent hydrolysis of carbonic acid as diprotic acid[11] with successive acid dissociation constants K_{a1} and K_{a2}, derive an effective Henry's law coefficient, k_{eff} (CO_2), for the dissolution of carbon dioxide in water defined by the relation

$$[CO_2(aq)_{tot}] = k_{\mathrm{eff}}(CO_2)p_{CO_2} \tag{7.8}$$

where $[CO_2(aq)_{tot}]$ and p_{CO_2} are the total concentration of CO_2 in the water and the equilibrium partial pressure of CO_2 in the air. Hence show that $k_{\mathrm{eff}}(CO_2) > k_H(CO_2)$.

Solution. When CO_2 dissolves in water, carbonic acid, that is $H_2CO_3(aq)$, is formed

$$CO_2(g) + H_2O(l) \rightleftarrows CO_2 \cdot H_2O(aq) \rightleftarrows H_2CO_3(aq) \tag{7.9}$$

This is followed by

$$H_2CO_3(aq) + H_2O(l) \rightleftarrows HCO_3^-(aq) + H_3O^+(aq) \tag{7.10}$$

and

$$HCO_3^-(aq) + H_2O(l) \rightleftarrows CO_3^{2-}(aq) + H_3O^+(aq) \tag{7.11}$$

Applying Henry's law to Reaction (7.9)

$$k_H(CO_2) = \frac{[H_2CO_3(aq)]}{p_{CO_2}} \tag{7.12}$$

For Reactions (7.10) and (7.11) at equilibrium we have, respectively,

$$K_{a1} = \frac{[H_3O^+(aq)][HCO_3^-(aq)]}{[H_2CO_3(aq)]} \tag{7.13}$$

and

$$K_{a2} = \frac{[H_3O^+(aq)][CO_3^{2-}(aq)]}{[HCO_3^-(aq)]} \tag{7.14}$$

Therefore, the concentrations of the three species in the water are

$$[H_2CO_3(aq)] = k_H(CO_2)p_{CO_2} \tag{7.15}$$

$$[HCO_3^-(aq)] = \frac{K_{a1}[H_2CO_3(aq)]}{[H_3O^+(aq)]} = \frac{K_{a1}k_H(CO_2)p_{CO_2}}{[H_3O^+(aq)]} \tag{7.16}$$

and

$$[CO_3^{2-}(aq)] = \frac{K_{a2}[HCO_3^-(aq)]}{[H_3O^+(aq)]} = \frac{K_{a1}K_{a2}k_H(CO_2)p_{CO_2}}{[H_3O^+(aq)]^2} \tag{7.17}$$

The total concentration of CO_2 dissolved in the water is

$$[CO_2(aq)]_{tot} = [H_2CO_3(aq)] + [HCO_3^-(aq)] + [CO_3^{-2}(aq)]$$

or

$$[CO_2(aq)]_{tot} = k_H(CO_2)p_{CO_2}\left(1 + \frac{K_{a1}}{[H_3O^+(aq)]} + \frac{K_{a1}K_{a2}}{[H_3O^+(aq)]^2}\right) \tag{7.18}$$

Comparing Eqs. (7.8) and (7.18), we see that

$$k_{eff}(CO_2) = k_H(CO_2)\left(1 + \frac{K_{a1}}{[H_3O^+(aq)]} + \frac{K_{a1}K_{a2}}{[H_3O^+(aq)]^2}\right) \tag{7.19}$$

where the values of K_{a1} and K_{a2} at 25°C and 1 atm are 4.4×10^{-7} and 4.7×10^{-11}, respectively. From Eq. (7.19) it follows that

$$k_{eff}(CO_2) > k_H(CO_2)$$

Equations (7.18) and (7.19) show that $[CO_2(aq)]_{tot}$ depends on $[H_3O^+(aq)]$, that is, on the pH of the solution. This is shown in Figure 7.5 where it can be seen that $k_{eff}(CO_2) \simeq k_H(CO_2)$ for pH ≤ 5, but for higher values of the pH both $k_{eff}(CO_2)$ and $[CO_2(aq)]_{tot}$ increase sharply.

Expressions for the effective Henry's law constants of SO_2, NH_3 (the only common basic gas in the atmosphere), nitric acid, and other gases can be derived in a similar way to that for CO_2 in Exercise 7.3.

Gases in the air that form strong electrolytes in solution, such as hydrogen iodide, nitric acid, and methane sulphonic acid, have Henry's law constants that exceed the largest values shown in Figure 7.4. Consequently, considerable amounts of these gases can dissolve in very small

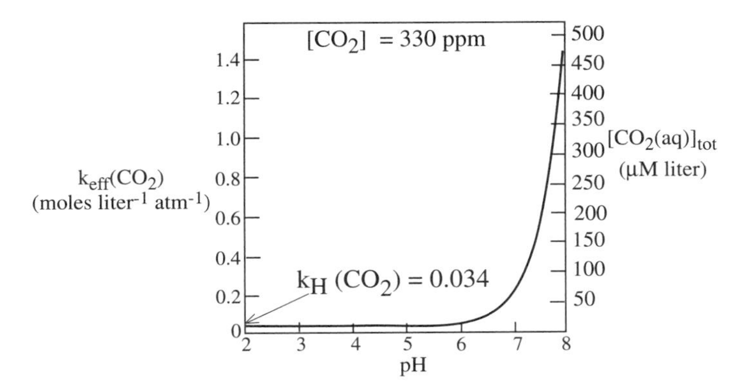

Figure 7.5. The effective Henry's law constant for CO_2, $k_{eff}(CO_2)$, and the equilibrium total dissolved CO_2, $[CO_2(aq)]_{tot}$, versus pH of the solution for a typical atmospheric partial pressure of $CO_2(g)$ of 330 ppmv. [Adapted from *Atmospheric Chemistry and Physics* by J. H. Seinfeld and S. N. Pandis. Copyright © 1998 John Wiley & Sons, Inc. Reprinted by permission of John Wiley & Sons, Inc.]

volumes of water (e.g., in the water of hydration of aerosols) to form very concentrated solutions.

7.4 Aqueous-phase chemical reactions

The relatively high concentrations of chemical species within cloud droplets, which derive initially from nucleation scavenging and the dissolution of gases, can lead to quite fast chemical reactions in the aqueous phase. To illustrate the basic principles involved, we will consider the important case of the conversion of dissolved sulfur dioxide to sulfate (i.e., the oxidation of S(IV) to S(VI)) in cloud water.[12]

a. Dissolution of SO_2

The dissolution of SO_2 can be handled in a similar way to CO_2 (see Exercise 7.3). When SO_2 dissolves in water, the bisulfite ion, $HSO_3^-(aq)$, and the sulfite ion, $SO_3^{2-}(aq)$, are formed

$$SO_2(g) + H_2O(l) \rightleftarrows SO_2 \cdot H_2O(aq) \tag{7.20}$$

$$SO_2 \cdot H_2O(aq) + H_2O(l) \rightleftarrows HSO_3^-(aq) + H_3O^+(aq) \tag{7.21}$$

$$HSO_3^-(aq) + H_2O(l) \rightleftarrows SO_3^{2-}(aq) + H_3O^+(aq) \tag{7.22}$$

Following the same steps as in Exercise 7.3, the following analogous expressions to Eqs. (7.18) and (7.19) are

$$[SO_2(aq)]_{tot} = [S(IV)]_{tot} = [SO_2 \cdot H_2O(aq)] + [HSO_3^-(aq)] + [SO_3^{2-}(aq)] \quad (7.23)$$

and

$$k_{eff}(SO_2) = k_H(SO_2)\left[1 + \frac{K_{a1}}{[H_3O^+(aq)]} + \frac{K_{a1}K_{a2}}{[H_3O^+(aq)]^2}\right] \quad (7.24)$$

where K_{a1} and K_{a2} now represent the successive acid dissociation constants of hydrated SO_2 (i.e., $SO_2 \cdot H_2O(aq)$, or $H_2SO_3(aq)$), namely, 1.3×10^{-2} and 6.2×10^{-8}, respectively.

Because of the much larger values of K_{a1} and K_{a2} for $H_2SO_3(aq)$ than for $H_2CO_3(aq)$, $k_{eff}(SO_2)$ increases much more quickly as the concentration of $H_3O^+(aq)$ decreases (i.e., the pH of the solution increases) than does $k_{eff}(CO_2)$. For example, $k_{eff}(SO_2)$ increases by almost seven orders of magnitude (!) as the pH increases from 1 to 8, compared to an increase of a factor of only about 44 in $k_{eff}(CO_2)$ for the same change in pH (see Fig. 7.5).

b. Oxidation of S(IV) to S(VI) in solution

As we have seen, the dissolution of SO_2 in an aqueous solution produces $SO_2 \cdot H_2O(aq)$ (or $H_2SO_3(aq)$), $HSO_3^-(aq)$, and $SO_3^{2-}(aq)$. In all three of these species, sulfur (S) is in the oxidation state 4 (i.e., S(IV)). This section is concerned with the subsequent (very fast) oxidation of the S(IV) species to S(VI), that is to $SO_4^{2-}(aq)$, in cloud water.

There are many aqueous-phase pathways for the oxidation of S(IV) to S(VI) in clouds. The more important ones are compared in Figure 7.6, where it can be seen that H_2O_2 (hydrogen peroxide) is the most important oxidant over a broad range of pH values, followed by O_3, some metal-catalyzed reactions, and NO_2. We will now consider briefly each of these oxidation paths.

Hydrogen peroxide is very soluble in water (see Fig. 7.4). It dissociates slightly to produce $HO_2^-(aq)$

$$H_2O_2(aq) + H_2O(l) \rightleftarrows HO_2^-(aq) + H_3O^+(aq)$$

but since it is a weak electrolyte this dissociation can usually be ignored. The conversion to $H_2SO_4(aq)$ proceeds as follows

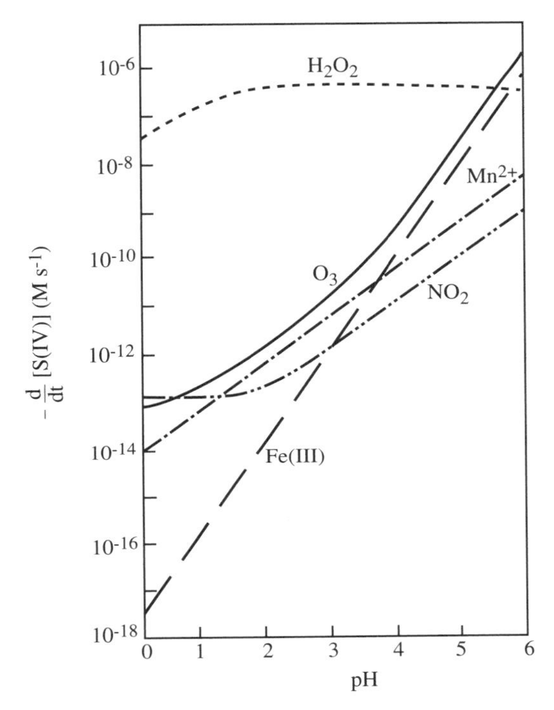

Figure 7.6. Comparison of the most important aqueous-phase pathways for the oxidation of S(IV) to S(VI). The ambient conditions assumed are $[SO_2(g)]$ = 5 ppbv, [NO(g)] = 1 ppbv, $[H_2O_2(g)]$ = 1 ppbv, $[O_3(g)]$ = 50 ppbv, [Fe(III) = 0.3 μM and [Mn(II)] = 0.03 μM. [From *Atmospheric Chemistry and Physics* by J. H. Seinfeld and S. N. Pandis. Copyright © 1998 John Wiley & Sons, Inc. Reprinted by permission of John Wiley & Sons, Inc.]

$$HSO_3^-(aq) + H_2O_2(aq) \rightleftarrows SO_2OOH^-(aq) + H_2O(l)$$

$$SO_2OOH^-(aq) + H_3O^+(aq) \rightarrow H_2SO_4(aq) + H_2O(l)$$

An empirical rate equation for the oxidation is

$$-\frac{d[S(IV)]}{dt} = \frac{k[H_3O^+(aq)][H_2O_2(aq)][HSO_3^-(aq)]}{1 + K[H_3O^+(aq)]} \qquad (7.25)$$

where $k = (7.5 \pm 1.6) \times 10^{-7}\,M^{-1}\,s^{-1}$ and $K = 13\,M^{-1}$ at 25°C. The relation (7.25) is plotted in Figure 7.6, where it can be seen that the oxidation rate is essentially independent of the pH of the solution. This is because the pH dependence of the S(IV) solubility and the reaction rate coefficient essentially offset each other. The reaction is so fast that $H_2O_2(g)$ and $SO_2(g)$ rarely coexist in liquid water clouds.

Next in importance to H_2O_2 as an oxidant for S(IV) in cloud droplets is ozone (Fig. 7.6). Indeed, at pH values in excess of about 5.5, O_3 rivals H_2O_2, even though ozone is only slightly soluble in water (see Fig. 7.4). The aqueous-phase reaction can be represented by

$$\mathrm{S(IV)} + \mathrm{O_3(aq)} \rightarrow \mathrm{S(VI)} + \mathrm{O_2(aq)}$$

Laboratory studies show that the rate of the reaction is given empirically by

$$-\frac{d}{dt}[\mathrm{S(IV)}] = \{k_0[\mathrm{SO_2 \cdot H_2O(aq)}] + k_1[\mathrm{HSO_3^-(aq)}] + k_2[\mathrm{SO_3^{2-}(aq)}]\}[\mathrm{O_3(aq)}]$$

where $k_0 = (2.4 \pm 1.1) \times 10^4\,\mathrm{M^{-1}\,s^{-1}}$, $k_1 = (3.7 \pm 0.7) \times 10^5\,\mathrm{M^{-1}\,s^{-1}}$ and $k_2 = (1.5 \pm 0.6) \times 10^9\,\mathrm{M^{-1}\,s^{-1}}$ at 25°C. The effective rate coefficient (k), defined by

$$-\frac{d}{dt}[\mathrm{S(IV)}] = k[\mathrm{A(aq)}][\mathrm{S(IV)}]$$

increases rapidly with increasing pH, from a value of about $10^5\,\mathrm{M^{-1}\,s^{-1}}$ at a pH of 2 to about $3 \times 10^7\,\mathrm{M^{-1}\,s^{-1}}$ at a pH of 6.

Oxygen (O_2) is quite soluble in water, but oxidation of S(IV) by dissolved O_2 is generally thought to be very slow in pure water. However, the oxidation rate is significantly enhanced by iron (Fe^{3+}) and manganese (Mn^{2+}), which are always present in trace amounts in cloud water. At pH values typical of cloud water, SO_2 is present mainly as HSO_3^-, and the catalyzed oxidation reaction can be represented by

$$2\mathrm{HSO_3^-(aq)} + \mathrm{O_2(aq)} + 2\mathrm{H_2O(l)} \xrightarrow{\mathrm{Fe^{3+},\,Mn^{2+}}} 2(\mathrm{SO_4^{2-}(aq)} + \mathrm{H_3O^+(aq)})$$

The oxidation rate is dependent on the pH of the solution (Fig. 7.6). There is some evidence that when Fe^{3+} and Mn^{2+} are both present the oxidation rate is 3 to 10 times faster than the sum of the two individual catalyzed rates shown in Figure 7.6. Consequently, the O_2 metal-catalyzed oxidation of S(IV) may exceed that of oxidation by O_3 and, at high pH values, it may be not much less than the H_2O_2 oxidation rate (see Fig. 7.6).

Nitrogen dioxide may react with dissolved SO_2 in water, in the form of $HSO_3^-(aq)$, to form S(VI)

$$2NO_2(aq) + HSO_3^-(aq) + 4H_2O(l) \rightarrow$$
$$3H_3O^+(aq) + 2NO_2^-(aq) + SO_4^{2-}(aq)$$

However, as shown in Figure 7.6, this reaction is slow; also the solubility of NO_2 in water is low (see Fig. 7.4).

Exercise 7.4. $SO_2(g)$ dissolves in a fog of water droplets to produce S(IV); the S(IV) then reacts with a dissolved species A to produce S(VI). The rate of depletion of S(IV) is given by

$$-\frac{d\,S(IV)}{dt} \equiv R = k[A(aq)][S(IV)]$$

where R is in moles of S(IV) consumed per second per liter of water, k is a rate coefficient (units: $M^{-1}\,s^{-1}$) and [A(aq)] and [S(IV)] both have units of moles per liter of water (i.e., M).

Derive an expression for the rate of depletion of S(IV) (in units of ppbv of $SO_2(g)$ depleted per hour) in terms of R, the liquid water content L of the cloud (in units of grams of water per cubic meter of air), the temperature T (in degrees K), and the universal gas constant in "chemical units," R_c^* (units: liter atm $deg^{-1}\,mol^{-1}$).

Assume that the number of molecules of S(IV) consumed per second is equal to the number of moles of $SO_2(g)$ depleted per second, and that the fog is at a pressure of 1 atm.

Solution. Let

R' = Rate of depletion of S(IV) in units of moles of SO_2 per sec per liter of *air*

Then,

$$R' = CR = C\,k[A(aq)][S(IV)]$$

where R is rate of depletion of S(IV) in moles per sec per liter of *water*, and C is the number of liter of fog water per liter of air. If the liquid water content of the fog is L grams per m^3 of air
1 m^3 of air contains L grams of liquid water

Therefore, 1 cm^3 of air contains $10^{-6}L$ grams of liquid water
$= 10^{-6}L\,cm^3$ of liquid water

Therefore, 1 liter of air contains $10^{-6}L$ liter of liquid water
Hence,

$$R' = (10^{-6}L)R = 10^{-6}L\,k[A(aq)][S(IV)] \quad (7.26)$$

where R' is in moles of S(IV) consumed per sec per liter of *air*.

For a gas X,

$$p_x = R_c^*[\mathrm{X}]T \tag{7.27}$$

where p_x is the partial pressure of X (in atm) and [X] the molarity.

1 liter of air contains [X] moles of gas X

Therefore, from Eq. (7.26),

$$R' = (10^{-6}L)R \quad \text{moles of S(IV) consumed per sec per [X] moles of } SO_2(g)$$

or

$$R' = \frac{(10^{-6}L)R}{[\mathrm{X}]} \quad \text{moles of S(IV) consumed per sec per mole of } SO_2(g) \tag{7.28}$$

From (7.27) and (7.28)

$$R' = \frac{(10^{-6}L)R\,R_c^*T}{p_{SO_2}} \quad \text{moles of S(IV) consumed per sec per mole of } SO_2(g) \tag{7.29}$$

where p_{SO_2} is the partial pressure of SO_2 (in atms).

But

moles of S(IV) consumed per sec
= moles of SO_2(g) depleted per sec

(This can be seen from Eqs. (7.20)–(7.22), since n moles of SO_2(g) produce $(n - x)$ moles of $SO_2 \cdot H_2O$(aq), $(x - y)$ moles of HSO_3^-(aq), and y moles of SO_3^{2-}(aq), or from Eq. (7.23), $[n - x) + (x - y) + y] = n$ moles of S(IV)].)
Therefore, from Eq. (7.29)

$$\begin{aligned} R' &= \frac{(10^{-6}L)R\,R_c^*T}{p_{SO_2}} \quad \text{moles of } SO_2(g) \text{ consumed per sec per mole of } SO_2(g) \text{ in the air} \\ &= \text{fractional rate of decrease in moles of } SO_2(g) \text{ per sec} \\ &= \text{fractional rate of decrease of partial pressure of } SO_2(g) \text{ per sec} \end{aligned}$$

Therefore,

$$\text{Fractional rate of decrease of partial pressure of } SO_2(g) \text{ per hour} = \frac{(10^{-6}L)R\,R_c^*T}{p_{SO_2}} 3{,}600$$

or

$$\begin{array}{ll}\text{Rate of decrease of partial} & (10^{-6}L)R\,R_c^*T\;3{,}600 \\ \text{pressure of } SO_2(g) & \text{(in atm per hour)}\end{array} \qquad (7.30)$$

Now,

$$\frac{\text{Partial pressure of } SO_2 \text{(in atm/hr)}}{\text{Total atmospheric pressure (in atm)}}$$
$$= \frac{\text{number concentration of } SO_2 \text{ molecules}}{\text{number concentration of all of the molecules in air}}$$
$$= \text{mixing ratio of } SO_2 \text{ by volume (as a fraction)}$$
$$= \text{mixing ratio of } SO_2 \text{ (in ppbv)} 10^{-9}$$

Therefore, since the total atmospheric pressure is 1 atm,

$$\begin{array}{ll}\text{Rate of decrease of} & \text{Rate of decrease of} \\ \text{partial pressure of } SO_2 = & \text{mixing ratio of } SO_2 \\ \text{(in atmos per hr)} & \text{(in ppbv per hr)} 10^{-9}\end{array} \qquad (7.31)$$

From (7.30) and (7.31),

$$\begin{array}{l}\text{Rate of decrease of mixing} \\ \text{ratio of } SO_2 \text{ (in pppv per hr)}\end{array} = 3.6 \times 10^6 LRR^*T$$

and since

$$\text{Rate of depletion of S(IV)} = \text{Rate of decrease of } SO_2$$

it follows that

$$\begin{array}{ll}\text{Rate of depletion of S(IV)} & \text{(in units of pppv of} \\ = 3.6 \times 10^6 LRR^*T & SO_2(g) \text{ per hour)}\end{array}$$

7.5 Precipitation scavenging

Precipitation scavenging refers to the removal of gases and aerosol particles (hereafter called *pollutants*) by liquid water and ice particles in clouds and by the precipitation elements (hereafter collectively called *hydrometeors*). Precipitation scavenging is crucially important for cleansing the atmosphere of pollutants, but it can also lead to severe impacts (in the form of acid rain) on the ground.

We have already discussed how aerosol particles may be incorporated into cloud water by acting as CCN, and the dissolution of gases into cloud

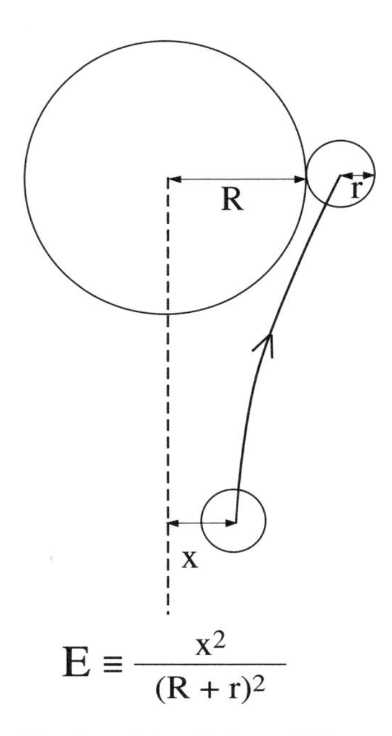

Figure 7.7. Definition of the inertial collision efficiency (E) of an aerosol particle of radius r with a hydrometeor of radius R. Here, x is the initial critical horizontal offset of the particle from the fall-line of the hydrometeor such that the aerosol particle just makes a glancing collision with the hydrometeor as it follows the airflow around the hydrometeor.

water. These two processes contribute to precipitation scavenging. Additional ways by which pollutants may be brought into contact with hydrometeors are through *diffusional* and *inertial* collisions. Diffusional collision refers to the diffusional migration of a pollutant through the air to a hydrometeor. The smaller the size of the pollutant the more efficient is diffusional collision. Therefore, small aerosol particles and gas molecules are collected relatively efficiently by diffusional collisions. Inertial collision refers to the collision of pollutants with hydrometeors which (by virtue of their larger sizes) fall relative to the pollutants. The smaller the size of a pollutant the more closely it will follow the airflow around a hydrometeor and therefore tend to avoid collision. Therefore, inertial collision is important only for larger aerosol particles (greater than a few micrometers).

Theoretical estimates of the diffusional and inertial *collision efficiencies* (see Fig. 7.7 for the definition of this term) for various sizes of aerosol

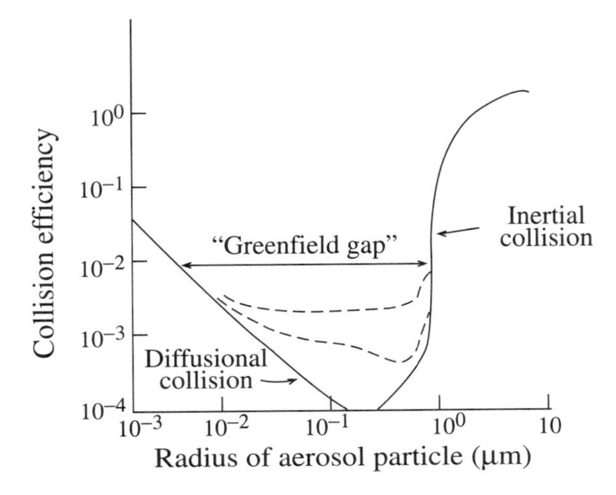

Figure 7.8. Estimates of collision efficiencies of aerosol particles with a raindrop 310 μm in radius by diffusional collision and inertial collision. The dashed lines show estimates of the contributions by phoretic and electrical effects. [Adapted from P. Wang et al., *J. Atmos. Sci.*, **35**, 1735 (1978).]

particles with a raindrop of a given size are shown in Figure 7.8. Since diffusional and inertial collisions are most efficient for capturing very small and relatively large sized pollutants, respectively, they leave a region of low collision efficiencies (sometimes called the *Greenfield gap*) for aerosol particles between ~0.1 and 1 μm in radius where neither of these two mechanisms is effective.

Laboratory and field measurements show reasonable agreement with the theoretical estimates shown in Figure 7.8 for the collection of aerosol particles greater than about 1 μm. But for submicrometer aerosol particles the measured collection efficiencies are generally higher than the theoretical estimates, particularly in the *Greenfield gap*. These higher collection efficiencies (indicated by the dashed lines in Fig. 7.8) may be due to phoretic effects (see Section 6.7), electrical effects, and the deliquescent growth of some aerosol particles (resulting in effective sizes that are larger than the dry sizes shown in Fig. 7.8).

The *collection* efficiency depends not only on the collision efficiency discussed above but also on the *coalescence* efficiency, which is the probability that the collision of an aerosol particle with a hydrometeor is followed by permanent collection. For convenience, the coalescence efficiency is often assumed to be 100%, although this may not always be the case.

7.6 Sources of sulfate in precipitation

The relative importance of nucleation scavenging, aqueous-phase chemical reactions, and below-cloud base precipitation scavenging to the amount of a particular chemical species that is in hydrometeors reaching the ground depends sensitively on the cloud and ambient conditions. This is illustrated for the case of sulfate (an important contributor to acid rain) by the following estimates from calculations based on the theoretical principles outlined in the previous sections.

Let us consider two quite different scenarios: a cloud situated in heavily polluted air (e.g., over the northeast United States), and a cloud situated in very clean air (e.g., over the southern oceans). Reasonable values of some of the main parameters required to make the calculations for these two scenarios are listed in Table 7.1. The results of the calculations are shown in Table 7.2. Considering first the cases where the collecting hydrometeors are raindrops (labeled scenarios 1 and 2 in Table 7.2), we see that in both cases (polluted and clean) nucleation scavenging and aqueous-phase chemical reactions dominate the concentration of sulfate in hydrometeors reaching the ground. For polluted air, aqueous-phase sulfate production dominates over nucleation scavenging (61% versus 37%). For clean air, the reverse is true (78% from nucleation scavenging and 18% from aqueous-phase chemical reactions). This difference is attributable to the $SO_2(g)$ concentration being one hundred times greater in the polluted air than in the clean air.

Table 7.1. *Values of parameters used in estimating the sulfur content of precipitation reaching the ground from a modest cumulonimbus cloud*

Parameter	Scenario 1 (polluted)	Scenario 2 (clean)
Concentration of SO_2 in air	10 ppb	0.1 ppb
Concentration of sulfate in air in which cloud forms	$3\ \mu g\ m^{-3}$	$1\ \mu g\ m^{-3}$
Rate of sulfate production in cloud water	7.9×10^{-8} mole s^{-1}	2.5×10^{-9} mole s^{-1}
Collection efficiency of sulfate by hydrometeors below cloud base	0.3	0.3
Cloud liquid water content	$0.5\ g\ m^{-3}$	$0.5\ g\ m^{-3}$
pH of cloud water	4.5	5.0

Table 7.2. *Estimates of the percentage contributions to the sulfate content of rain for scenarios 1 and 2 described in Table 7.1 assuming that the cloud particles and the hydrometeors are liquid water, and assuming the collecting hydrometeors have an ice-phase origin (indicated by 1* and 2*)*

	Percentage Contribution to Sulfate Content			
Scenario	Nucleation Scavenging	Aqueous-Phase Chemical Reactions	Below-Cloud Base Precipitation Scavenging	Total Sulfate Concentrations in Rain at Ground Level (M)
1 (polluted)	37	61	2	1.4×10^{-4}
2 (clean)	78	18	4	2.2×10^{-5}
1* (polluted)	27	71	2	1.2×10^{-4}
2* (clean)	76	18	6	1.4×10^{-5}

Two other scenarios are listed in Table 7.2 (labeled 1* and 2*), which are the same polluted and clean cases shown in Table 7.1, but for the case where the collecting hydrometeors are taken to originate as ice (e.g., snow crystals or hailstones). In the case of clean air, this does not have much effect on the relative contributions to the sulfate concentration in hydrometeors reaching the ground. However, if it is assumed that the hydrometeors originate as ice in the polluted air, there is significant increase in the *relative* contribution of the aqueous-phase chemical reactions. This is because in this case the hydrometeors do not originate on a CCN, therefore the contribution from nucleation scavenging is reduced.

7.7 Chemical composition of rainwater

The pH of pure water is 7. The pH of rainwater under very clean atmospheric conditions is 5.6. This is due in part to the absorption of CO_2 in rainwater and the formation of carbonic acid (see Exercise 7.3 and pages 89 and 90 of *Basic Physical Chemistry for the Atmospheric Sciences* by P. V. Hobbs, Cambridge University Press, 2000). However, as shown in Figure 7.9, the pH of rainwater in polluted air can be significantly lower than 5.6, giving rise to what is known as *acid rain*. This is due to the incorporation of gaseous and particulate pollutants in rain by the mechanisms

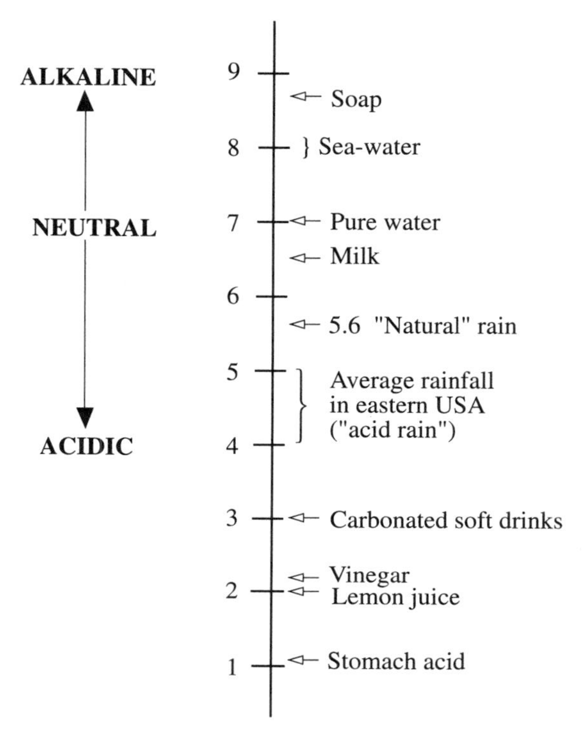

Figure 7.9. The pH of some common substances compared to that of pure water, "natural" rain, and "acid rain."

discussed in Sections 7.1–7.5. In Section 7.6 we discussed the contribution of sulfates to the acidification of rain. However, species other than sulfate contribute to acid rain (Table 7.3). For example, in Pasadena, California, there is more nitrate than sulfate in rain, and the ammonium ion (NH_4^+) is present in relatively high concentrations. Poker Flat, Alaska, although in a remote location, has rain and snow with an average pH of 4.96, with sulfate the main contributor. In the eastern United States, sulfuric acid and nitric acid contribute about 65% and 30%, respectively, to the acidity. Due to the variability of both the natural sources of chemicals and meteorological parameters, the contributors to acid rain and the pH of rainwater, are highly variable both in remote and urban areas.

Table 7.3. *Mean chemical composition (in µeq per liter) of rain in a remote clean area (Poker Flat, Alaska[a]) and in a polluted urban area (Pasadena, California[b])*

Species	Poker Flat, Alaska (Mean for 1980–1981)	Pasadena, California (Mean for 1984–1987)
SO_4^{2-}	10.2	18.7
NO_3^-	2.4	23.8
Cl^-	4.8	21.7
Mg^{2+}	0.5	5.1
Na^+	2.1	19.8
K^+	1.2	1.1
Ca^{2+}	0.5	7.6
NH_4^+	2.0	18.8
pH	4.96	4.72

[a] From J. Galloway et al., *J. Geophys. Res.*, **89**, 1447 (1982). Copyright by the American Geophysical Union.
[b] From J. H. Seinfeld and S. N. Pandis, *Atmospheric Chemistry and Physics.* John Wiley, New York, 1998.

7.8 Production of aerosols by clouds

We have seen that, due to aqueous-phase chemical reactions in cloud droplets, particles released from evaporating clouds may be larger and have different chemical compositions than the CCN on which the cloud droplets formed. In this section, we describe another way in which clouds can affect atmospheric aerosol.

Figure 7.10 shows some measurements of humidity and Aitken nucleus concentrations obtained from an aircraft as it flew across the upper regions of five marine cumulus clouds. The clouds themselves are indicated by the shaded areas in Figure 7.10. It can be seen that surrounding each cloud are regions of enhanced humidity and enhanced Aitken nucleus concentrations. The enhanced humidity is to be expected, since clouds moisten the surrounding air. The enhanced aerosol concentrations can be explained as follows.

In the same way as small water droplets can form by the combination of water molecules in air that is highly supersaturated, under appropriate conditions the molecules of two gases can combine to form aerosol particles (called *homogeneous, bimolecular nucleation*). The condi-

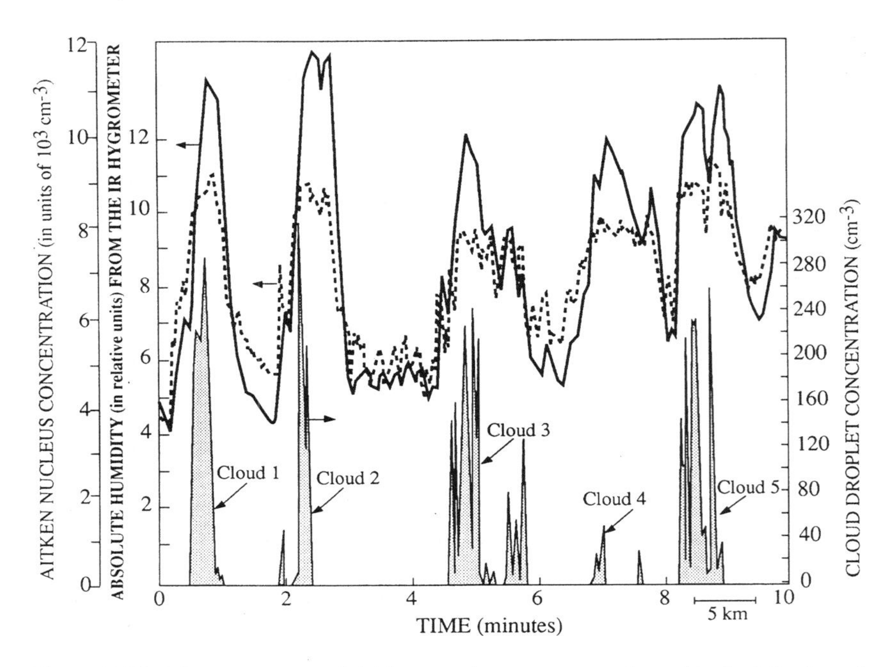

Figure 7.10. Measurements of Aitken nucleus concentrations (dashed line) and humidity (solid line) across five small marine cumulus clouds (shaded regions) indicated by the cloud droplet concentrations. [From L. F. Radke and P. V. Hobbs, *J. Atmos. Sci.*, **48**, 1190 (1991).]

tions that favor the formation of new particles by this process are high concentrations of appropriate gases, low ambient concentrations of aerosols (which might otherwise provide surfaces onto which the gases condense, rather than condensing as new particles), and low temperatures (which encourage condensation). These conditions are sometimes met in the outflow regions of clouds, as indicated schematically in Figure 7.11.

Figure 7.11 depicts a cumulus cloud over the ocean. Air from the marine boundary layer is transported into the base of the cloud in the updraft region. This air contains aerosol particles of various sizes and chemical compositions, as well as trace gases (e.g., SO_2, DMS, H_2SO_4, O_3). Inside the cloud, the total aerosol particle number concentration and the total surface area of the unactivated aerosol particles decrease with height above cloud base due to in-cloud removal mechanisms. These removal mechanisms include Brownian diffusion and

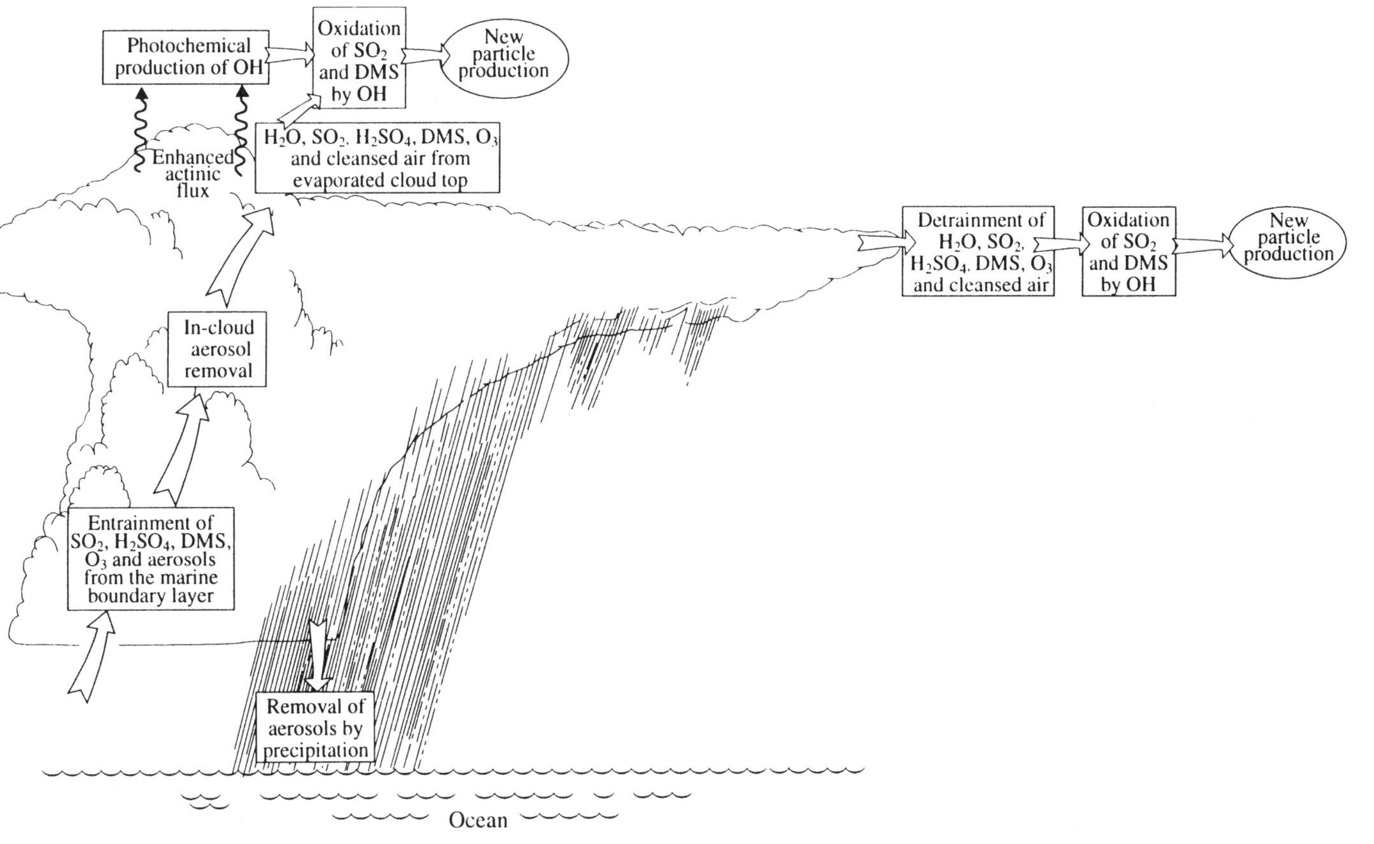

Figure 7.11. Schematic diagram illustrating a conceptual model for new particle production near marine convective clouds. [From K. Perry and P. V. Hobbs, *J. Geophys. Res.*, **99**, 22813 (1994). Copyright © by the American Geophysical Union.]

diffusiophoresis of cloud interstitial aerosol to hydrometeors, collisional scavenging by cloud droplets and ice crystals, and subsequent removal by precipitation. A significant, but variable, fraction of the SO_2 and H_2SO_4 will be removed from the air in the cloud by aqueous sulfate production. However, due to their extremely low solubilities in water, both DMS and O_3 will be essentially unaffected by the presence of cloud droplets.

Cloudy air is detrained into the environment, after a significant fraction of the aerosol particles have been removed by the cloud. This detrainment will occur in the anvil outflow region and above cloud top. The detrained air raises the relative humidity and vents SO_2, H_2SO_4, DMS, and O_3 to the near-cloud environment. In the clear air, O_3 can be photolyzed to form the OH radical. The OH then oxidizes SO_2 to form H_2SO_4. Under conditions of high relative humidity, low total aerosol surface area, low temperatures, and high SO_2 concentrations in the outflow from the cloud, significant concentrations of new particles can be produced by homogeneous-bimolecular nucleation of sulfuric acid solution droplets.

The aerosol nucleation process may be accelerated above cloud tops due to the high actinic flux at this location, which can be elevated by up to a factor of three above thick clouds due to multiple reflections of solar radiation. The enhanced actinic flux can photolyze O_3 and produce OH concentrations significantly higher than ambient values via Reactions (5.17a) and (5.17b), respectively. The high OH concentrations create high H_2SO_4 concentrations, which accelerate new particle production above cloud top.

The processes illustrated in Figure 7.11 may act on a large scale to supply large numbers of Aitken nuclei to the upper troposphere in the tropics and subtropics and also, perhaps, to the subtropical marine boundary layer (MBL). In this scenario (Fig. 7.12), air containing DMS from the oceans is transported in large tropical convective clouds to the upper troposphere. New aerosol production occurs in the outflow regions of the clouds by the homogeneous-bimolecular nucleation of sulfuric acid solution droplets. These aerosols are then transported poleward by the upper branch of the Hadley cell circulation, and then subside in the subtropics. It has even been suggested that during this transport some of the particles may grow large enough (by further condensation, coagulation, and cloud processing) to provide a significant source of CCN (large enough to be activated in stratiform clouds) to the subtropical marine boundary layer.

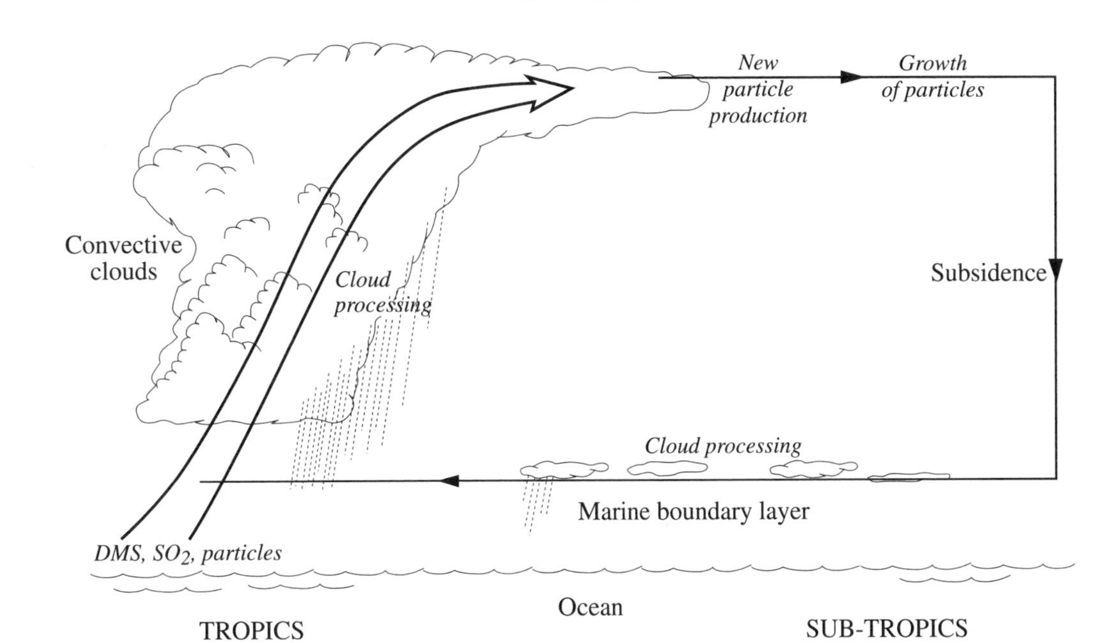

Figure 7.12. The Hadley cell scenario for transporting aerosols between the tropics and sub-tropics. [Adapted from F. Raes et al., in *Dimethylsulphide: Oceans, Atmosphere and Climate*, Ed. G. Restelli and G. Angeletti, Kluwer Academic Publishers (1993), with permission from Kluwer Academic Publishers.]

Exercises

See Exercise 1(w) and Exercises 33–40 in Appendix I.

Notes

1 The theory of water vapor condensation in the absence of aerosols (which requires supersaturations in excess of several hundreds of percent, which never occur in the Earth's atmosphere) is given in Section 2.8 of *Basic Physical Chemistry for the Atmospheric Sciences* by P. V. Hobbs, Cambridge University Press, New York (2000).

2 A surface is said to be perfectly wettable (*hydrophilic*) if water spreads out on it as a horizontal film. It is completely unwettable (*hydrophobic*) if water forms spherical droplets on its surface.

3 Lord Kelvin 1st Baron (William Thomson) (1924–1907). Scottish mathematician and physicist. Entered Glasgow University at age 11. At 22, became Professor of Natural Philosophy at the same university. Carried out incomparable work in thermodynamics, electricity, and hydrodynamics.

4 For a derivation of Kelvin's equation see Section 2.8 of *Basic Physical Chemistry for the Atmospheric Sciences* by P. V. Hobbs, Cambridge University Press, New York (2000).

5 See Section 4.4 of *Basic Physical Chemistry for the Atmospheric Sciences* by P. V. Hobbs, Cambridge University Press, New York (2000) for the derivation of Eq. (7.2), which is known as *Raoult's law*.

6 H. Köhler, (1888–1982). Swedish meteorologist. Former Director of the Meteorological Observatory, University of Uppsala.

7 The discussion in this paragraph implies that CCN play a dominant role in determining cloud microstructures. However, cloud droplets are also affected by the entrainment of drier ambient air into the cloud, and by collisions between droplets. The reader is referred to books on cloud physics for more detailed discussions of the many processes that can affect cloud microstructures.

8 Some of the principles involved in the effects of CCN on the reflectivity of clouds are illustrated by Exercise 33 in Appendix 1.

9 See Chapter 4 of *Basic Physical Chemistry for the Atmospheric Sciences* by P. V. Hobbs, Cambridge University Press, New York (2000) for a more extensive discussion of the basic principles of solution chemistry.

10 Strictly speaking, Henry's law holds only for variations in solution composition over the range for which the solution is ideally dilute (i.e., obeys Raoult's law, Eq. 7.2). Also, the concentration of a gas in a droplet will be given by Henry's law only if other processes are not rate limiting. Other processes include transfer of the gas through the air to the surface of the droplet, transfer of the gas across air-water interface, and mixing of the gas within the droplet. The first of these processes is very rapid (~1 ms). The transfer of relevant atmospheric gases across the air-water interface is also quite rapid (~0.05 to 1 s). Mixing of a gas within a droplet, to achieve the uniform concentration assumed by Henry's law, occurs in a few tenths of a second. Since all three processes occur on time scales much less than the lifetime of a cloud droplet, they are generally not rate limiting. The characteristic times for hydrolysis ionization in droplets is virtually instantaneously; chemical reactions in droplets are also very fast.

11 See Chapter 5 of *Basic Physical Chemistry for the Atmospheric Sciences* by P. V. Hobbs, Cambridge University Press, New York (2000) for discussions of hydrolysis, polyprotic acids, and other aspects of the chemistry of acids and bases.

12 For a more detailed discussion of cloud chemistry, the reader is referred to *Atmospheric Chemistry and Physics* by J. H. Seinfeld and S. N. Pandis, John Wiley and Sons Inc., New York (1998).

8
Tropospheric chemical cycles

The reservoirs of chemical species in the Earth system are the solid Earth, the hydrosphere (oceans and fresh water), the cryosphere (ice and snow), the biosphere, and the atmosphere. Chemical species can be transferred between these reservoirs. We have already seen that such transfers played crucial roles in the evolution of Earth's atmosphere. Since, under steady-state conditions, a chemical species cannot accumulate indefinitely in any one of the reservoirs, there must be continual cycling of species through the various reservoirs. This is termed *biogeochemical cycling*.

In this chapter we will consider briefly the tropospheric portions of the biogeochemical cycles of carbon, nitrogen, and sulfur. We have chosen these cycles not only because they are of considerable importance in atmospheric chemistry but because the productivity of many terrestrial and aquatic organisms depends on the availability of these elements (as well as on oxygen and hydrogen). We will be concerned with relatively rapid interchanges involving the atmosphere and other reservoirs (generally the oceans and the biosphere). For simplicity, we will assume the various cycles are independent, although in the atmosphere they can interact with each other and are coupled in the stoichiometry of living matter and biochemical processes. A few words of caution are needed. Many aspects of global chemical cycles are not well understood, and the magnitudes of the various emission fluxes are not necessarily equivalent to the importance of the species, since atmospheric residence times must also be taken into account.

8.1 Carbon cycle

The exchanges of carbon between the atmosphere and biosphere are shown in Figure 8.1. The fluxes of reactive carbon-containing gases are

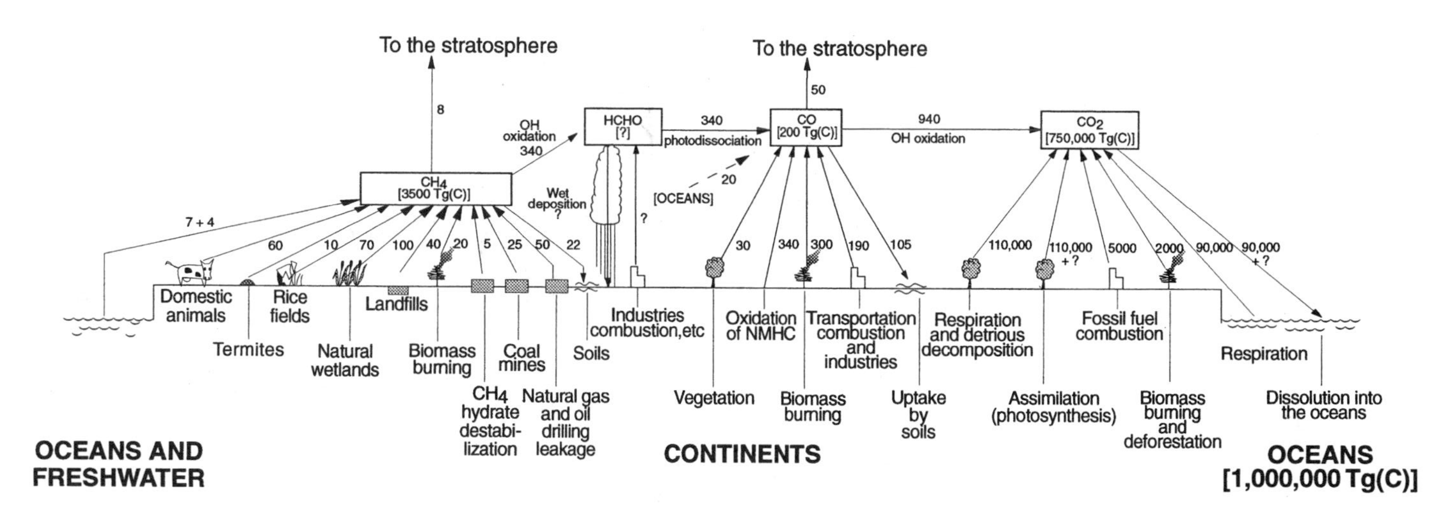

Figure 8.1. The principal sources and sinks of carbon-containing gases in the atmosphere. The numbers alongside the arrows are estimates of average annual fluxes in Tg(C) per year; various degrees of uncertainty, some quite large, are associated with all of the fluxes. The numbers in square brackets are the total amounts of the species in the atmosphere.

shown on the left side of the figure; the fluxes involving CO_2, which is the primary carbon-containing gas in the atmosphere but is relatively unreactive, are shown on the right side of Figure 8.1. Although the fluxes of the reactive gases are small compared to that of CO_2, they are associated with several trace species of great importance in atmospheric chemistry (CH_4, HCHO, and CO).

The major sources of CH_4 for the atmosphere are believed to be bacterial degradations of organic matter in rice fields and wetlands, fermentation in the stomachs of domestic animals, and leakage from natural gas and oil drilling (several other lesser sources are shown in Fig. 8.1). The sinks for CH_4 are better known than the sources. They are oxidation by OH to formaldehyde (HCHO), uptake by soils, and destruction in the stratosphere. Although the strengths of the various sources of CH_4 shown in Figure 8.1 are quite uncertain, it can be seen that anthropogenic sources make the major contribution (>60%).

The amounts of carbon stored in the atmosphere and biosphere are tiny compared to those stored in the oceans and solid Earth (see Table 1.2). However, as far as the natural carbon cycle is concerned, the main interchanges with the atmosphere on relatively short time scales occur with the biosphere. For example, from the definition of residence time (Eq. (2.4)), and from the amount of CO_2 in the atmosphere and its effluxes via photosynthesis and dissolution into the ocean (Figure 8.1), the atmospheric residence of CO_2 considering both photosynthesis and dissolution into the ocean is

$$\sim \frac{750,000}{110,000 + 90,000} = 3.8\,\mathrm{a}$$

and considering just photosynthesis it is

$$\sim \frac{750,000}{110,000} = 6.8\,\mathrm{a}$$

By comparison the residence time for dissolved organic carbon in the deep oceans is ~3,000 a, and for carbon stored in the solid Earth it is ~10^8 a. Consequently, for the exchange of carbon over short time periods (<100 a), the atmosphere-biosphere can be considered as essentially isolated from the rest of the carbon cycle.

Problem 8.1. Using the information given in Figure 8.1, estimate (a) the residence time of CH_4 in the troposphere with respect to influxes, and (b) the annual percentage increase in the amount (and therefore the concentration) of CH_4 in the troposphere.

Solution. (a) From Figure 8.1 we see that the magnitude of the tropospheric reservoir of CH_4 is 3,500 $Tg(C)$. By adding the influxes of CH_4 shown in Figure 8.1, the total influx is found to be 391 $Tg(C)$ a^{-1}. From Eq. (2.4),

$$\begin{aligned}\text{residence time of CH}_4\text{ in the troposphere} &= \frac{\text{magnitude of tropospheric reservoir}}{\text{total influx to troposphere}}\\ &= \frac{3,500}{391}\\ &\simeq 9\,\text{a}\end{aligned}$$

(b) The total efflux of CH_4 from the troposphere is, from Figure 8.1, 370 $Tg(C)\,a^{-1}$. Therefore,

$$\text{total influx of CH}_4 - \text{total efflux of CH}_4 = 391 - 370 = 21\,Tg(C)\text{a}^{-1}$$

The annual rate of increase of CH_4 in the atmosphere should be 21 $Tg(C)\,a^{-1}$, which, expressed as a percentage is $(21 \times 100)/3{,}500 = 0.60\%$. The measured increase in the concentration of CH_4 in the 1980s was ~0.9%. However, the close agreement between this value and our calculated value is deceptive. This is because the fluxes shown in Figure 8.1 were estimated as follows. First the annual fluxes were estimated, since they are known much better than the influxes. Then, using measurements of the average annual rate of increase in atmospheric CH_4 concentrations, the annual increase in the CH_4 troposphere reservoir was determined. The total influx of CH_4 to the troposphere was then set equal to the sum of the total efflux and the annual increase; the total influx was then distributed among the various sources based on best estimates of their relative magnitudes. Thus, the information shown in Figure 8.1 has been constrained to produce the observed annual rate of increase in atmospheric CH_4!

Formaldehyde (HCHO) is the most reactive gas shown in Figure 8.1. Over the oceans the only major source of HCHO is the oxidation of CH_4 by OH. Over land, there are many direct sources of HCHO (industries, automobiles, biomass burning), but their magnitudes are not well known. Formaldehyde is removed from the atmosphere by precipitation. It is also photodissociated into CO by sunlight, which gives HCHO a residence time of ~5 h in midlatitudes when averaged over a full day.

The major sources of CO are oxidation of CH_4 via HCHO, biomass burning, oxidation of hydrocarbons (mainly natural), and fossil-fuel combustion. The main sink is OH oxidation to CO_2, with smaller sinks due to consumption by soils, and transport to the stratosphere. It can be seen from the estimates of the fluxes given in Figure 8.1 that anthropogenic emissions make a dominant contribution (~50%) to the CO budget.

The fluxes of CO_2 into and out of the atmosphere are dominated by plant respiration and assimilation (photosynthesis). However, these influxes and effluxes are very closely in balance. Consequently, the fluxes of CO_2 into the atmosphere from anthropogenic sources (primarily fossil fuel consumption and biomass burning) have produced significant perturbations to the atmospheric CO_2 budget. From the magnitudes of the fluxes shown in Figure 8.1, the influx of CO_2 to the atmosphere exceeds the efflux by 7,940 $Tg(C)\,a^{-1}$. Based on the measured increase of ~1.8 ppmv a^{-1} in the atmospheric concentrations of CO_2 in the 1980s (see Fig. 1.1), the increase in the atmospheric CO_2 reservoir was ~3,000 $Tg(C)$ a^{-1}. Since this exceeds the estimated net influx of CO_2 by 7,940 – 3,000 = 4,940 $Tg(C)\,a^{-1}$, some adjustments must be made to the fluxes shown in Figure 8.1. The difference of 4,940 $Tg(C)\,a^{-1}$ is probably taken up by the oceans and the terrestrial ecosystem (indicated by the question marks in Fig. 8.1).

If the magnitude of the atmospheric CO_2 reservoir (750,000 $Tg(C)$) is divided by the total efflux of CO_2 [200,000 $Tg(C)\,a^{-1}$], a residence time of ~4 a is obtained. Thus, on average, it takes only a few years before a CO_2 molecule in the atmosphere is taken up by plants or dissolved in the oceans. However, as can be seen from Figure 8.1, there are no real sinks for CO_2, rather it is circulated between various reservoirs (atmosphere, ocean, biota). Therefore, 4 a is not the time that it would take for atmospheric CO_2 concentrations to adjust to a new equilibrium if the sources and sinks were changed. It is the latter adjustment time that is relevant to "greenhouse" warming caused by increasing atmospheric CO_2 concentrations. This adjustment time (50 to 200 a), which is the one given in Table 2.1, is determined by the slow exchange of carbon between the surface of the ocean and the deep ocean.

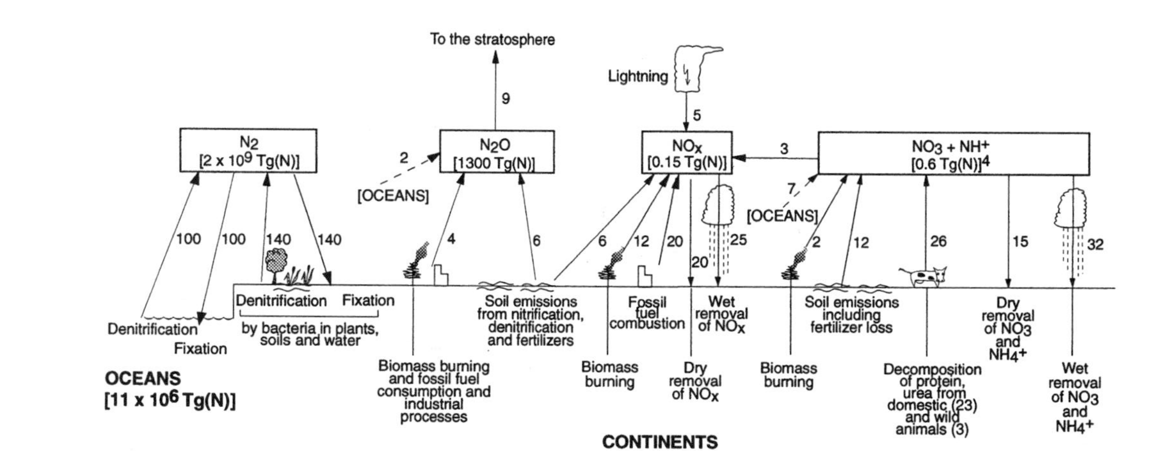

Figure 8.2. As for Fig. 8.1 but for nitrogen-containing gases. Fluxes are in Tg(N) per year.

8.2 Nitrogen cycle

Nitrogen gas (N_2) constitutes more that 99.9999% of the nitrogen present in the atmosphere, and N_2O makes up more than 99% of the rest of the nitrogen. The other nitrogen species in the atmosphere (see Table 3.1) are therefore present in very low concentrations, but they are of crucial importance in atmospheric chemistry. For example, NH_3 is the only basic gas in the atmosphere. Therefore, it is solely responsible for neutralizing acids produced by the oxidation of SO_2 and NO_2; the ammonium salts of sulfuric and nitric acid so formed become atmospheric aerosols. Nitric oxide and NO_2 play important roles in both tropospheric and stratospheric chemistry.

All of the nitrogen-containing gases in the air are involved in biological nitrogen fixation and denitrification. *Fixation* refers to the reduction and incorporation of nitrogen from the atmosphere into living biomass, which is accomplished by certain bacteria that are equipped with a special enzyme system for this task; the usual product is NH_3. The term *fixed nitrogen* refers to nitrogen contained in chemical compounds that can be used by plants and microorganisms. Under aerobic (i.e., oxygen-rich) conditions, other specialized bacteria can oxidize ammonia to nitrite and then to nitrate; this is called *nitrification*. Most plants use nitrate taken through their roots to satisfy their nitrogen needs. Some of the nitrate undergoes bacterial reduction to N_2 and N_2O (termed *denitrification*), which returns fixed nitrogen from the biosphere to the atmosphere. In this case, nitrate acts as the oxidizing agent; therefore, denitrification generally occurs under anaerobic conditions, where oxygen is unavailable. Fixed nitrogen can also be returned from plants to the atmosphere via N_2O. Biomass burning returns fixed nitrogen to the atmosphere as N_2, NH_3, N_2O, and NO_x.

The principal sources and sinks of nitrogen-containing species in the atmosphere are shown in Figure 8.2. Assuming that fixation and nitrification of N_2 are in approximate balance, the main sources of nitrogen-containing species are biogenic emissions from the Earth and the oceans (NH_3, N_2O, and NO_x), decomposition of proteins and urea from animals (NH_3), biomass burning and fossil fuel consumption (NO_x, NH_3, and N_2), and lightning (NO_x). The main sinks are wet removal by precipitation (NH_3 and NO_x as NO_3^-, dry deposition (NO_x and NH_3), and the chemical breakdown of N_2O in the stratosphere. Anthropogenic sources of NH_3, N_2O, and NO_x (from fossil fuel consumption, biomass burning, and agricultural nitrate fertilization) are appreciable. Therefore, they may be

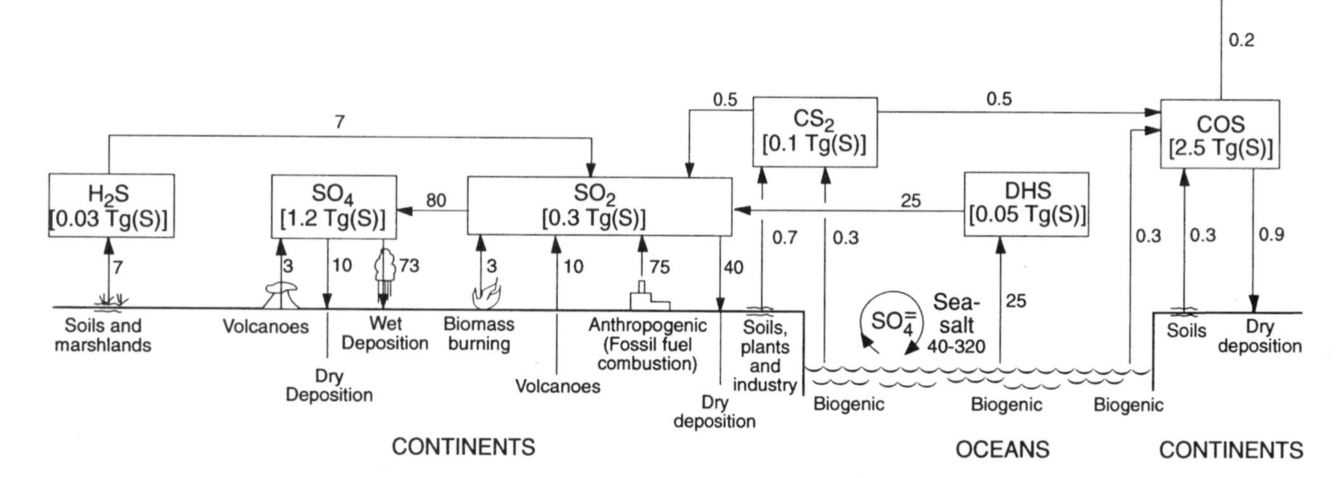

Figure 8.3. As for Fig. 8.1 but for sulfur-containing species in the troposphere. Fluxes are in Tg(S) per year. For clarity, wet and dry removal are shown only over the continents, although they occur also over the oceans.

causing significant perturbations in the budgets of nitrogen species in the atmosphere (see Exercise 42 in Appendix I).

8.3 Sulfur cycle

The most important reduced sulfur gases in the air are H_2S, DMS, COS, and CS_2. Their main natural sources are biogenic reactions in soils, marshland, and plants (for H_2S, DMS, COS, and CS_2), and biogenic reactions in the ocean due primarily to phytoplankton (for DMS, COS and CS_2). When these gases are released into the oxygen-rich atmosphere they are oxidized to SO_2, and then over 65% of the SO_2 is oxidized to $SO_4^=$ (the remainder of the SO_2 is removed by dry deposition). Estimates of the fluxes of these natural emissions of sulfur gases, and their transformations to SO_2 and $SO_4^=$ are given in Figure 8.3.

It can be seen from Figure 8.3 that DMS dominates the emissions from the ocean. (An enormous amount of sulfate is ejected into the air from the oceans in the form of sea salt. However, since these are relatively large particles, they are quickly recycled back to the ocean. Therefore, they do not have a significant effect on the global sulfur cycle.) Most of the sulfides in the air are oxidized rapidly by the OH radical; therefore, their residence times are only days to a week. An important exception is COS, which is very stable in the troposphere and, as a result, has a relatively long residence time (~2 a) and a large and relatively uniform concentration (~0.5 pppv, compared to ~0.4 ppbv for DMS, ~0.2 ppbv for H_2S and SO_2, and ~0.05 pptv for CS_2). Consequently, COS is the most abundant sulfur compound in the troposphere; however, because it is relatively long-lived, it is generally ignored in tropospheric chemistry. Its relatively long residence time enables it to be gradually mixed into the stratosphere, where, converted by UV radiation, it is the dominant source of sulfate aerosol.

The flux of anthropogenic emissions of sulfur to the atmosphere is known quite well; it is ~78 $Tg(S)\,a^{-1}$, which is similar to (actually somewhat greater than) estimates of the natural emissions of reduced sulfur gases to the atmosphere (Fig. 8.3). Therefore, the global sulfur budget is significantly affected by anthropogenic emissions. Anthropogenic emissions of sulfur are almost entirely in the form of SO_2. The main sources are the burning of coal and the smelting of sulfide ores. The main mechanisms for removing sulfur from the atmosphere are wet and dry deposition. For example, of the 80 $Tg(S)\,a^{-1}$ of SO_2 that are oxidized to $SO_4^=$, about 70 $Tg(S)\,a^{-1}$ occur in clouds, which are subsequently wet deposited, and the remainder is oxidized by gas-phase reactions, which is dry deposited (Fig. 8.3).

Exercises

See Exercises 1(x)–(z) and Exercises 41–45 in Appendix I.

9
Air pollution

We have seen in Chapter 8 that, even on a global scale, anthropogenic emissions can cause significant perturbations to the budgets of certain trace chemical species in the atmosphere. In urban and industrialized areas, anthropogenic emissions can become so large that the concentrations of various undesirable chemical species (called *pollutants*) cause significant deterioration in air quality and visibility, and can pose threats to human health. Severe air pollution episodes occur when the rates of emissions or formation of pollutants greatly exceed the rates at which the pollutants are dispersed or destroyed by winds or by vertical transport (e.g., in the presence of a capping temperature inversion) or removed from the atmosphere (e.g., by precipitation or by chemical Reactions).

9.1 Sources of anthropogenic pollutants

Combustion (in power plants, smelters, automobiles, and of wood, etc.) is the largest source of air pollutants. On a global scale fossil-fuel combustion is the major source of CO, CO_2, NO_x, and SO_2 (see Figs. 8.1–8.3). Many other pollutants are released into the air by combustion. For example, about 15% of the total emissions of hydrocarbons are from anthropogenic sources; this is because the most common fuels are hydrocarbon compounds (oil, natural gas, coal, and wood). Ideal (or complete) oxidation (or combustion) of hydrocarbon fuel yields only CO_2 and H_2O. However, for a given quantity of fuel, a precise amount of oxygen is required for complete combustion.

Problem 9.1. Determine the ratio of the mass of dry air to the mass of isooctane (C_8H_{18}) – called the *air–fuel ratio* – for complete combustion.

Solution. The balanced chemical equation for the complete combustion of C_8H_{18} is

$$C_8H_{18} + 12.5O_2 \rightarrow 8CO_2 + 9H_2O$$

Therefore, for complete combustion, 1 mole of C_8H_{18} reacts with 12.5 moles of O_2. Or, since the molecular weights of C_8H_{18} and O_2 are 114 and 32, respectively, 114 grams of C_8H_{18} reacts with (12.5×32) or 400 grams of O_2. We now need to calculate what mass of air contains 400 grams of oxygen. The percentage of oxygen in air *by volume* (or by number of molecules) is 20.95% (see Table 3.1). Since the apparent molecular weight of dry air is 28.97, the percentage of oxygen in air *by mass* is $20.95 \times (32/28.97) \simeq 23\%$. Therefore, the mass of air containing 400 grams of oxygen is $400/0.23 \simeq 1{,}700$ g. Hence, for complete combustion, 114 grams of C_8H_{18} reacts with about 1,700 grams of air. Therefore, the air–fuel ratio for complete combustion is $1{,}700/114 \simeq 15$.

Since 1981, gasoline-powered internal combustion engines in the United States have used an oxygen sensor in the exhaust system and a computer-controlled fuel flow that keeps the air–fuel mixture stoichiometric (i.e., perfect for complete combustion) within a few percent. This system, together with the catalyst, makes their emissions quite small. Older automobiles tended to run with a fuel-rich mixture (i.e., a mixture that contains less than the required amount of air for complete combustion), which has different exhaust chemistry from the stoichiometric. For example, suppose the 12.5 moles of O_2 (in Problem 9.1) is reduced to 11.25 moles. Then, 9 moles of H_2O are produced together with 5.5 moles of CO_2 and 2.5 moles of CO. More than one-third of the carbon in the emissions is now the highly poisonous gas CO. Gas mileage is reduced because the fuel is not completely burned, but because the engine is pumping less air, peak power can actually increase. Even modern cars operate in this "fuel-rich mode" for almost 1 min when cold starting, and for several seconds when driven at full throttle. Therefore, despite the fact that in most countries automobiles use only a small percentage of the total fuel burned, they produce a large fraction of the CO. In the United States alone, ~100 Tg of CO are produced each year.

Carbon monoxide is relatively unreactive, but it does bind strongly to hemoglobin (the iron-containing protein that carries oxygen in blood). Thus, relatively small quantities of CO restrict the transport of oxygen by blood. The concentration of CO in city traffic often reaches 50 ppmv, and it can rise to 140 ppmv (compared to ~0.05 ppmv in clean, unpolluted air). Even these CO concentrations are low compared to that in inhaled cigarette smoke, which is ~10,000 ppmv!

If an automobile has a broken catalyst, it will emit significant amounts of unburned (and partially burned) fuel hydrocarbon (HC). Ignition system failure, or overly lean operation, causes misfiring, which increases the HC emissions (many of which are toxic and carcinogenic). Measurements of motor vehicle exhausts on highways show that half of the emissions of CO and HC come from less than 10% of old (or poorly maintained) vehicles.

Combustion of conventional fuels also produces NO. This is because the high temperatures permit oxidation of atmospheric molecular nitrogen to nitric oxide (referred to as *thermal* NO). At temperatures below ~4,500 K, the reactions are

$$O_2 + M \rightleftarrows 2O + M \qquad (9.2a)$$

$$O + N_2 \rightleftarrows NO + N \qquad (9.2b)$$

$$N + O_2 \rightleftarrows NO + O \qquad (9.2c)$$

Because of the strong temperature dependence of Reactions (9.2a) and (9.2b), thermal NO formation is very temperature dependent. Under equilibrium conditions, these reactions produce maximum NO concentrations at ≥3,500 K (although equilibrium is not attained in engines). As the combustion gases cool rapidly to ambient temperatures the rates of the reverse reactions are drastically reduced, so that the NO concentration is "frozen" at the high temperature value. An additional rapid source for NO production during combustion is the oxidation of nitrogen-containing compounds in the fuel (*fuel* NO).

As we will see in Section 9.2, NO_x emissions from automobiles play a key role in the formation of photochemical smog. However, over large geographic areas, power stations and industries are generally larger sources of NO_x than automobiles.

Fuel combustion, particularly of coal, dominates the emissions of sulfur oxides, which are mainly SO_2. Heavy metal smelters (e.g., Ni, Cu, Zn, Pb, and Ag) can be large local sources of SO_2.

There are also lower temperature sources of air pollutants. For example, leakages of hydrocarbons from natural gas lines, organics from the evaporation of solvents, nitrogen gases from fertilizers, and chlorofluorocarbons (CFC) from refrigerants and the electronic industry.

As discussed in Sections 6.7(d) and 6.3(b), anthropogenic activities also emit large quantities of aerosols into the atmosphere, both directly and through g-to-p conversion. For particles $\geq 5\,\mu m$ diameter, human activities worldwide are estimated to produce ~15% of natural emissions, with industrial processes, fuel combustion, and g-to-p conversion accounting for ~80% of the anthropogenic emissions. However, in urban areas anthropogenic sources are much more important. For particles $<5\,\mu m$ diameter, human activities produce ~20% of natural emissions, with g-to-p conversion accounting for ~90% of the anthropogenic emissions.

The *emission factor* is a convenient way of comparing chemical emissions from various processes. It is defined as the mass of a chemical emitted into the atmosphere per unit mass of material processed (e.g., mass of SO_2 emitted per unit mass of coal burned). Some emission factors are given in Table 9.1.

9.2 Some atmospheric effects of air pollution

a. Classical (or London) smog

The term *smog* derives from *smo*ke and fo*g*; it was originally coined to refer to heavily polluted air that can form in cities (generally in winter under calm, moist conditions) due to the emissions of sulfur dioxide and aerosols from the burning of fossil fuels (primarily coal and oil). Prior to the introduction of air pollution abatement laws in the latter part of the twentieth century, many large cities in Europe and North America regularly suffered from severe smogs. The London smogs were sufficiently notorious that such pollution became known as *London smog*.[1]

In the London (or *classical*) type of smog, the aerosols in the smoke serve as nuclei on which fog droplets form. Then, SO_2 gas absorbs into the fog droplets where it is oxidized to form sulfuric acid (see Section 7.4). In the infamous London smog of 1952, which lasted for four days and was implicated in the deaths of ~4,000 people, SO_2 reached peak concentrations of ~0.7 ppm (compared to typical annual mean concentrations of ~0.1 ppm in polluted cities with large coal usage), and the peak aerosol smoke concentrations were ~$1.7\,mg\,m^{-3}$. Interestingly, there is no reason to believe that these concentrations would in themselves have caused fatalities. Therefore, the deaths must have been due to synergistic action coupled, perhaps, with the effect of the low temperatures that existed at the time of the smog.

Table 9.1. *Some emission factors of gases and particles from anthropogenic sources*

Type of Emission	Emission Factor (in 1990s)
(a) Coal-fired Electric Power Plants	
NO_x (as NO_2) from plants without pollution controls	10 g per kg of coal burned
SO_2 from plants without pollution controls, using coal containing 2% sulfur	38 g per kg of coal burned
United States Environmental Protection Agency's New Source Performance Standards for NO_x (as NO_2) or SO_2	7.5 g per kg of coal burned
Total particles from plant without pollution controls, burning pulverized coal containing 10% ash	50 g per kg of coal burned
Total particles from plants with electrostatic precipitator emission control units, burning pulverized coal containing 10% ash	0.4 g per kg
(b) Copper Smelters	
SO_2 from smelter without pollution controls, using ore with 25% Cu and 30% sulfur	530 g per kg of ore and 2,120 g per kg of Cu
(c) Automobiles	
CO in uncontrolled exhaust[a]	275 g per kg of gasoline used
CO with pollution controls required in the United States[a]	13 g per kg of gasoline used
Hydrocarbons (C) in uncontrolled exhaust[a]	24 g per kg of gasoline used
Hydrocarbons (C) with pollution controls required in the United States[a]	1.6 g per kg of gasoline used
NO_x (as NO_2) in uncontrolled exhaust[a]	14 g per kg of gasoline used
NO_x (as NO_2) with pollution controls required in the United States[a]	4 g per kg of gasoline used
(d) Biomass Burning (of Forests in Brazil)	
CO_2 from flaming forest fires	913 g of carbon per kg of carbon burned
CO from flaming forest fires	60 g of carbon per kg of carbon burned
Total particles from flaming forest fires	10 g of carbon per kg of carbon burned

[a] Based on average speed of 40 km h^{-1} (25 mph) in urban areas.

b. Photochemical (or Los Angeles) smog

During the second half of the twentieth century emissions from automobiles became increasingly important as a source of pollutants in many large cities and urban areas. When subjected to sunlight, and stagnant meteorological conditions, the combination of chemical species in such strongly polluted urban air can lead to *photochemical* (or *Los Angeles–type*) *smog*. These smogs are characterized by high concentrations of a large variety of pollutants, such as nitrogen oxides, ozone (which damages plants and materials), carbon monoxide (a poisonous gas), hydrocarbons, aldehydes (and other materials that are eye irritants), and sometimes sulfuric acid. The chemical reactions that lead to this type of smog are extremely complex, and still not completely understood. Next, we give an outline of some of the major chemical reactions that are thought to be involved.

Photochemical smogs result from the interactions of a variety of organic pollutants (e.g., hydrocarbons such as ethylene and butane) with oxides of nitrogen. The reactions start with one we have already discussed, namely, the photolysis of NO_2

$$NO_2 + h\nu \xrightarrow{j} NO + O \tag{9.3}$$

Ozone is then formed very quickly by

$$O + O_2 + M \xrightarrow{k_1} O_3 + M \tag{9.4}$$

where M is most likely N_2 or O_2. However, not much O_3 is formed directly by Reaction (9.4) since it is depleted by the rapid reaction

$$O_3 + NO \xrightarrow{k_2} NO_2 + O_2 \tag{9.5}$$

which regenerates NO_2. If there were no other reactions, (9.3)–(9.5) would lead to a steady-state concentration of O_3 given by

$$[O_3] = \frac{j[NO_2]}{k_2[NO]} \tag{9.6}$$

Equation (9.6), called the *Leighton relationship*, predicts ozone concentrations in urban polluted air of only ~0.03 ppmv, whereas typical values are well above this concentration and can reach 0.5 ppmv. Therefore, other chemical reactions must be involved. Most effective would be reactions that oxidize NO to NO_2 without consuming O_3, since this would allow the ozone to build up during the day. The OH radical can initiate a chain reaction that can act in this way by attacking the hydrocarbon pollutants in urban air, for example

$$OH + CH_4 \rightarrow H_2O + CH_3 \tag{9.7}$$

or

$$OH + CO \rightarrow H + CO_2 \tag{9.8}$$

or

$$OH + CH_3CHO \rightarrow H_2O + CH_3CO \tag{9.9}$$

The resulting radicals, CH_3 in Reaction (9.7), H in (9.8), and CH_3CO in (9.9), then become involved in reactions that oxidize NO to NO_2 and regenerate OH. For example, CH_3 from Reaction (9.7) can initiate the following series of reactions

$$CH_3 + O_2 \rightarrow CH_3O_2 \tag{9.10a}$$

$$CH_3O_2 + NO \rightarrow CH_3O + NO_2 \tag{9.10b}$$

$$CH_3O + O_2 \rightarrow HCHO + HO_2 \tag{9.10c}$$

$$HO_2 + NO \rightarrow NO_2 + OH \tag{9.10d}$$

The net effect of Reactions (9.7) and (9.10) is

$$CH_4 + 2O_2 + 2NO \rightarrow H_2O + 2NO_2 + HCHO \tag{9.11}$$

Thus, in this case, the hydrocarbon CH_4 oxidizes NO to NO_2 without consuming O_3. Reaction (9.11) produces formaldehyde (HCHO), which is an eye irritant. Note that this chain reaction scheme, involving the odd hydrogen radicals OH and HO_2, does not remove odd hydrogen.

Similarly, the acetyl radical (CH_3CO) from Reaction (9.9) is involved in a series of reactions leading to the methyl radical CH_3 and the peroxyacetyl radical (CH_3COO_2). The methyl radical oxides NO through Reactions (9.10) and the peroxyacetyl radical reacts with nitrogen dioxide

$$CH_3COO_2 + NO_2 \rightarrow CH_3COO_2NO_2 \tag{9.12}$$

The chemical species on the right hand side of Reaction (9.12) is the vapor of a colorless and dangerously explosive liquid called peroxyacetyl nitrate (PAN), which is an important component of photochemical smogs and another major eye irritant.[2] Other alkenes (e.g., propane)

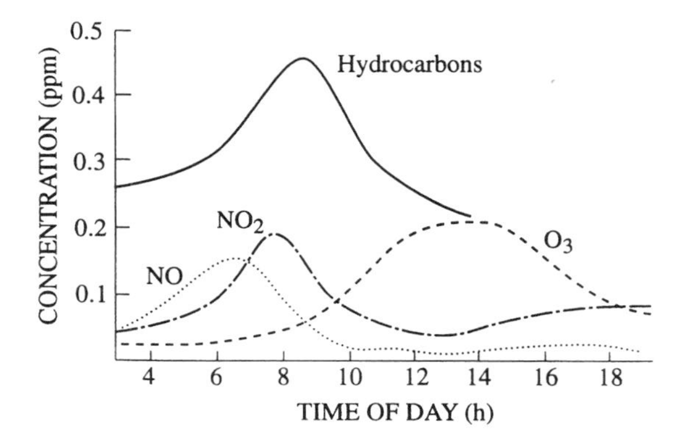

Figure 9.1. Typical variations during the course of a day of some important pollutants in photochemical smogs in Los Angeles. [Adapted from P. A. Leighton, *Photochemistry of Air Pollution*, Academic Press (1961).]

oxidize NO to NO_2 without consuming O_3 and regenerate OH, and can do so faster than the reactions discussed earlier.

Shown in Figure 9.1 are variations through the course of a day in the concentrations of some of the major components of photochemical smogs in Los Angeles.

c. *Visibility reduction*

Perhaps the most noticeable effect of air pollution is reduction in visibility or, to be more precise, a reduction in *meteorological range* (R). Meteorological range is defined as the distance from an observer at which an ideal black object just disappears when viewed against the horizon sky in daytime. It depends on the composition of the air (particularly its aerosol content), the amount and distribution of light, and the visual acuity of the observer. Particles and gases in the air reduce R by scattering and absorbing light from the object being viewed, as well as scattering extraneous light to the eye from the sun and sky. For a homogeneous atmosphere, the meteorological range is given by the *Koschmeider equation*

$$R = \frac{3.9}{b_e} \tag{9.13}$$

where b_e is equal to the sum of the scattering (b_s) and absorption (b_a) coefficients due to aerosols and gases.

For an aerosol-free atmosphere, b_e is due solely to molecular scattering and absorption by gases. In this case, at sea level and at a wavelength of 0.520 μm, $b_e = 1.3 \times 10^{-5}\,\text{m}^{-1}$. Substituting this value of b_e into (9.13) yields a value for R of $3 \times 10^5\,\text{m}^{-1}$ or 300 km; this meteorological range can be used as a benchmark for evaluating the effects of aerosols on visibility reduction.

Problem 9.2. Scattering by gaseous molecules (i.e., *Rayleigh scattering*) is proportional to the density of the air. If the meteorological range due to such scattering is 300 km at sea level (1,012 mb), what is it at a pressure level of 500 mb? Assume that the temperatures at 1,013 and 500 mb are 15 and –21°C, respectively.

Solution. Let the meteorological ranges at 1,012 and 500 mb due to Raleigh scattering alone be $R_{1,012}$ and R_{500}, respectively, and the air densities at 1,012 and 500 mb be $\rho_{1,012}$ and ρ_{500}, respectively. Since the scattering coefficient is proportional to the density of the air, it follows from Eq. (9.13) that the meteorological range is inversely proportional to the density. Therefore,

$$\frac{R_{1,012}}{R_{500}} = \frac{\rho_{500}}{\rho_{1,012}} \tag{9.14}$$

From the ideal gas equation

$$p = R\rho T \tag{9.15}$$

where p, ρ, and T are the pressure, density and temperature (in K) of the gas, and R is a constant. From Eqs. (9.14) and (9.15)

$$\frac{R_{1,012}}{R_{500}} = \frac{p_{500}/T_{500}}{p_{1,012}/T_{1,012}}$$

where the subscripts indicate the pressure levels for p and T. Therefore,

$$\frac{R_{1,012}}{R_{500}} = \frac{500/252}{1,012/288} = 0.56$$

$$\therefore \quad R_{500} = \frac{R_{1,012}}{0.56}$$

$$= \frac{300}{0.56}\,\text{km}$$

$$R_{500} = 536\,\text{km}$$

That is, the meteorological range at pressure level of 500 mb (about 5.5 km) due to Raleigh scattering alone is 536 km!

In heavily polluted urban/industrial areas, median values of R at midday are about 30 km, and in clean rural areas R is about 150 km. The main contributors to visibility reduction are aerosols with diameters from about 0.1 to 1 μm (i.e., similar to the wavelength of visible light) consisting of $(NH_4)_2$, $SO_4^=$, NH_4NO_3, and organics. Most of the extinction is due to the scattering of light by these aerosols, but in very polluted air absorption of light by elemental (or black) carbon can contribute up to 50% to b_e.

Nitrogen dioxide is the only atmospheric gas that absorbs sufficiently at visible wavelengths to make a significant contribution to b_e. In heavily polluted air, absorption by NO_2 can contribute up to about 6% to b_e. Since NO_2 absorbs mainly blue light, it can give certain industrial plumes a reddish-brown appearance. However, the brown coloration of thick industrial smogs is due primarily to aerosol scattering.

d. Regional and global pollution

The effects of anthropogenic pollution now extend to regional and global scales. For example, Europe, Russia, the northeastern United States, and large areas of southeastern Asia, are regularly covered by enormous palls of polluted air that reduce visibility significantly, give rise to wet and dry acid deposition, soil and erode buildings and other materials, and have deleterious effects on human health, animals and plants.

The fact that pollutants can be transported over large distances is well illustrated by air pollution in the Arctic, which can sometimes become as severe as that in industrial areas. Such pollution episodes are known as *arctic haze*. The pollutants originate from fossil-fuel combustion, smelting, and other industrial processes in northern Europe and Russia. The pollutants are transported to the Arctic by synoptic-scale airflow patterns, primarily from December to April. Since the arctic atmosphere is generally stably stratified during this time of the year, vertical mixing is limited; also, precipitation is rare so that wet removal processes are weak. Consequently, the pollutants can be transported over large distances with relatively little dilution. A major contributor to arctic haze is SO_2, which is converted to sulfate particles over the long transport distances. Glacial records indicate that arctic air pollution has increased markedly since the 1950s, paralleling the increases in SO_2 and NO_x emissions in Europe. Interestingly, analysis of time series of ice cores from Greenland has revealed unusually high lead concentrations during the period ~500 B.C. to 300 A.D. This is attributed to Greek and Roman lead and silver mining and smelting activities, which apparently polluted large regions of the Northern Hemisphere. However, cumulative lead deposits in the Greenland ice during these eight centuries was only ~15% of that caused by the widespread use of lead additives in gasoline from ~1930–1995.

The effects of pollution on a global scale are well illustrated by the world-wide increase in CO_2 concentrations since the beginning of the twentieth century (see Fig. 1.1). Other trace gases associated with pollution (e.g., CO, CH_4, N_2O, and various CFCs) also show increasing concentrations worldwide. CO_2, CH_4, N_2O, and CFCs are *greenhouse gases* that can cause global warming (see Section 4.7). As we will see in Section 10.2, the CFC can also cause serious problems in the stratosphere.

Exercises

See Exercises 1(aa)–(cc) and Exercise 46 in Appendix I.

Notes

1 When the author was a boy in London in the 1950s, smogs were often so thick that visibility was restricted to a few feet and buses had to be guided by pedestrians! However, smogs occurred in cities long before the modern industrial era. As early as 1661 the English diarist John Evelyn noted the effects of industrial emissions (and, no doubt, domestic wood burning) on the health of plants and people. He suggested that industries be placed outside of towns and that they be equipped with tall chimneys to disperse the smoke.

2 The discovery of PAN is an interesting chemical detective story. Before it was identified chemically, it was called "compound X," and some people attributed all of the ills from air pollution in Los Angeles to this mysterious compound. Eventually it was discovered that a new family of oxidants (peroxyacyl nitrates) was produced in strongly polluted urban air and that compound X was a member of this family.

10
Stratospheric chemistry

The stratosphere extends from the tropopause to a height of ~50 km above the Earth's surface (see Fig. 3.1). The height of the tropopause increases from ~10 km over the poles to ~17 km over the equator, but it varies with the seasons and meteorological conditions. In passing across the tropopause, from the troposphere to the stratosphere, there is generally an abrupt change in concentrations of many of the trace constituents in the air. For example, water vapor decreases, and ozone (O_3) concentrations often increase by an order of magnitude, within the first few kilometers above the tropopause. The strong vertical gradients across the tropopause reflect the fact that there is very little vertical mixing between the relatively moist, ozone-poor troposphere and the dry, ozone-rich stratosphere. Within the stratosphere the air is generally neutral or stable with respect to vertical motions. Also, the removal of aerosols and trace gases by precipitation, which is a powerful cleansing mechanism in the troposphere, is generally absent in the stratosphere. Consequently, materials that enter the stratosphere (e.g., volcanic effluents, anthropogenic chemicals that diffuse across the tropopause or are carried across the tropopause by strong updrafts in deep thunderstorms, and effluents from aircraft) can remain there for long periods of time, often as stratified layers.

In this chapter, three topics of particular interest in stratospheric chemistry are considered: unperturbed (i.e., natural) stratospheric O_3, anthropogenic perturbations to stratospheric O_3, and sulfur in the stratosphere. Our emphasis is on chemical processes. However, it should be kept in mind that in the atmosphere chemical, physical, and dynamical processes are often intimately entwined.

10.1 Unperturbed stratospheric ozone

a. Introductory remarks

Ozone in the stratosphere is of great importance for the following reasons:

- Ozone forms a protective shield that reduces the intensity of UV radiation (with wavelengths between 0.23 and 0.32 μm)[1,2] from the Sun that reaches the Earth's surface.
- Ozone determines the vertical profile of temperature in the stratosphere because of heating resulting from the UV absorption.
- Ozone is involved in many stratospheric chemical reactions.

In 1881 Hartley[3] measured the UV radiation reaching the Earth's surface and found a sharp cut off at $\lambda = 0.30\,\mu$m; he correctly attributed this to O_3 in the stratosphere. Subsequent ground-based UV measurements as a function of the Sun's elevation, and the first measurements of the concentrations of stratospheric O_3 by balloons in the 1930s placed the maximum O_3 concentration in the lower part of the stratosphere.

Shown in Figures 10.1 and 10.2 are the results of more recent measurements of O_3. The presence of an *ozone layer* between heights of ~15 and 30 km is very clear. However, the O_3 layer is highly variable: its height and intensity change with latitude, season, and meteorological conditions. Figure 10.1 includes the total *ozone column abundance*, which is the total amount of O_3 in a unit cross-sectional area from the surface of the Earth to the top of the atmosphere. The SI units for this quantity are kg m^{-2} or molecules m^{-2}, but it is common to express them as Dobson[4] units (DU). One DU is the thickness, in units of hundredths of a millimeter, that the total O_3 column would occupy at 0°C and 1 atm. Remarkably, the Earth's protective O_3 layer corresponds to only about 300 DU; that is, if all the ozone in the atmosphere were brought to a pressure of 1 atm at 0°C, it would form a layer only about 3 mm deep.

The largest column densities of O_3 in the northern hemisphere occur in polar latitudes in spring (Fig. 10.2(a)); in the southern hemisphere the spring maximum is at midlatitudes. Since O_3 is produced by photochemical reactions, the production is a maximum in the stratosphere over the tropics. The peaks in concentrations at polar and midlatitudes must be attributed to meridional transport of O_3 away from the equator, although at any given point in the atmosphere the balance between the production and loss of O_3, and its flux divergence, will determine the O_3

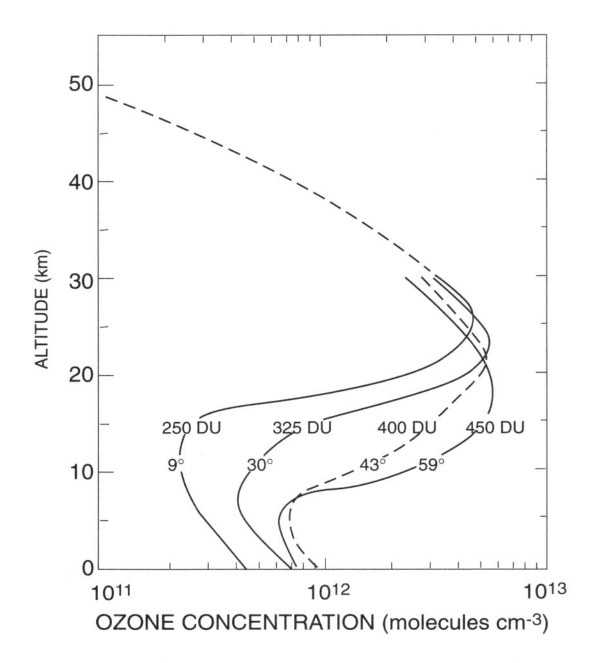

Figure 10.1. Mean vertical distributions of ozone concentrations based on measurements at different latitudes (given in degrees). Note the increase in the total ozone column abundance (given in DU) with increasing latitude. [Adapted from G. Brasseur and S. Solomon, *Aeronomy of the Middle Atmosphere*, D. Reidel Pub. Co., 215 (1984), with permission from Kluwer Academic Publishers.]

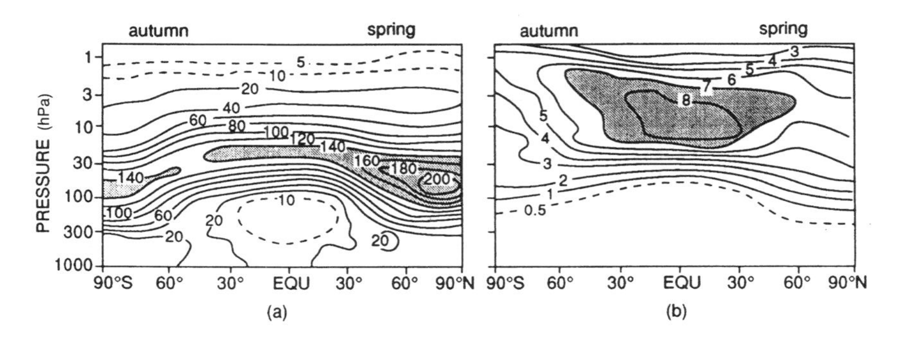

Figure 10.2. Meridional cross sections of ozone in the atmosphere. (a) Expressed as partial pressure in units of nbar. (b) Expressed as mixing ratio in units of ppmb. [From H. U. Dütsch, *Neujahrsblatt der Naturforschenden Gesellschaft in Zürich auf das Jahr 1980*, 182. Stück.]

concentration at that point. It is clear from Figure 10.1 that much of the change in the total column abundance of O_3 is due to differences in the profiles below 20 km.

Since 1960 remote sensing measurements from satellites have provided a wealth of information on the global distribution of O_3 and the variations in its vertical profiles and column abundances.

b. Chapman's theory

In 1930 Chapman[5] proposed a simple chemical scheme for maintaining steady-state concentrations of O_3 in an "oxygen-only" stratosphere. The reactions were the dissociation of O_2 by solar UV radiation (at $\lambda < 0.242\,\mu\text{m}$)

$$O_2 + h\nu \xrightarrow{j_a} O + O \tag{10.1}$$

the reaction of atomic oxygen and molecular oxygen to form O_3

$$O + O_2 + M \xrightarrow{k_b} O_3 + M \tag{10.2}$$

(where M represents N_2 or O_2), the photodissociation of O_3 (which occurs for λ from $0.20\,\mu\text{m}$ to $0.30\,\mu\text{m}$)

$$O_3 + h\nu \xrightarrow{j_c} O_2 + O \tag{10.3}$$

and the combination of atomic oxygen and O_3 to form O_2

$$O + O_3 \xrightarrow{k_d} O_2 + O_2 \tag{10.4}$$

Reactions (10.1)–(10.4) are called the *Chapman Reactions*.

Exercise 10.1. Assuming photochemical equilibrium, derive an expression for the rate of change of the concentration of O_3 (n_3) in terms of the concentrations of atomic oxygen (n_1), O_2 (n_2), and the inert molecule (n_M), and the rate coefficients for Reactions (10.1)–(10.4).

Solution. The source of O_3, and therefore the positive contribution to dn_3/dt, is provided by Reaction (10.2), which generates O_3 at a rate of $k_b n_1 n_2 n_M$. The sinks of O_3, and therefore the negative contributions to dn_3/dt, are provided by Reactions (10.3) and (10.4), which remove O_3 at rates of $j_c n_3$ and $k_d n_1 n_3$, respectively. Therefore,

$$\frac{dn_3}{dt} = k_b n_1 n_2 n_M - j_c n_3 - k_d n_1 n_3 \tag{10.5}$$

(Expressions for dn_1/dt and dn_2/dt may be derived in a similar way – see Exercise 47 in Appendix I)

Exercise 10.2. During the daytime in the stratosphere a steady-state concentration of atomic oxygen may be assumed. Use this fact to derive an expression for dn_3/dt in terms of n_1, n_2, n_3, k_d, and j_a. Assuming that the removal of atomic oxygen by Reaction (10.4) is small, and (at 30 km) $j_c n_3 \gg j_a n_2$, derive an expression for the steady-state concentration of atomic oxygen in the stratosphere at 30 km during the daytime in terms of n_2, n_3, n_M, k_b, and j_c. Hence, derive an expression for dn_3/dt in terms of $n_2, n_3, n_M, j_a, j_c, k_b$, and k_d.

Solution. From Reactions (10.1)–(10.4) we have

$$\frac{dn_1}{dt} = 2j_a n_2 - k_b n_1 n_2 n_M + j_c n_3 - k_d n_1 n_3 \tag{10.6}$$

If atomic oxygen is in steady state, $dn_1/dt = 0$ and

$$2j_a n_2 + j_c n_3 = k_b n_1 n_2 n_M + k_d n_1 n_3 \tag{10.7}$$

From Eqs. (10.5) and (10.7)

$$\frac{dn_3}{dt} = 2j_a n_2 - 2k_d n_1 n_3 \tag{10.8}$$

If the removal of atomic oxygen by Reaction (10.4) is small, we have from Eq. (10.6)

$$\frac{dn_1}{dt} \simeq 2j_a n_2 - k_b n_1 n_2 n_M + j_c n_3 \tag{10.9}$$

Therefore, since dn_1/dt and $j_c n_3 \gg 2j_a n_2$,

$$n_1 \simeq \frac{j_c n_3}{k_b n_2 n_M} \tag{10.10}$$

Substituting Eq. (10.10) into Eq. (10.8) yields

$$\frac{dn_3}{dt} = 2j_a n_2 - \frac{2k_d j_c n_3^2}{k_b n_2 n_M} \tag{10.11}$$

The Chapman reactions reproduce some of the broad features of the vertical distribution of O_3 in the stratosphere. For example, they predict that O_3 concentrations should reach maximum values at a height of ~25 km. Indeed, prior to 1964 it was generally believed that the Chapman theory provided an adequate description of stratospheric ozone chemistry. This is now known not to be the case. For example, although the Chapman reactions predict about the right shape for the vertical profile of O_3, they overpredict the concentrations of O_3. Also, model calculations based on the Chapman reactions predict that the global rate of production of O_3 in spring is 4.86×10^{31} molecules s^{-1} (of which 0.06×10^{31} molecules s^{-1} are transported to the troposphere). However, the loss of "odd oxygen"[6] is only 0.89×10^{31} molecules s^{-1}. This leaves a net 3.91×10^{31} molecules s^{-1}, which would double atmospheric O_3 concentrations in just two weeks.[7] Since O_3 concentrations are not increasing, there must be important sinks of odd oxygen in the stratosphere in addition to Reaction (10.4). This brings us to the subject of catalytic chemical cycles in the stratosphere.

c. Catalytic chemical cycles

Most of the catalytic reactions that have been proposed for the removal of stratospheric odd oxygen are of the form

$$X + O_3 \rightarrow XO + O_2 \quad (10.12a)$$

$$XO + O \rightarrow X + O_2 \quad (10.12b)$$

$$\text{Net:} \quad O + O_3 \rightarrow 2O_2 \quad (10.13)$$

where X represents the catalyst and XO the intermediate. Provided that Reactions (10.12) are fast, Reaction (10.13) can proceed much faster than Reaction (10.4). Also, since X is consumed in Reaction (10.12a) but regenerated in Reaction (10.12b), and provided there is no appreciable sink for X, just a few molecules of X have the potential to eliminate indefinite numbers of O_3 molecules and atomic oxygen.

In the natural (i.e., anthropogenically undisturbed) stratosphere the most important contenders for the catalyst X are H, OH, NO, and Cl. For example, in the case of NO

$$NO + O_3 \rightarrow NO_2 + O_2 \quad (10.14a)$$

$$NO_2 + O \rightarrow NO + O_2 \quad (10.14b)$$

$$\text{Net:} \quad O + O_3 \rightarrow 2O_2 \quad (10.15)$$

At a temperature of –53°C (which is typical of the stratosphere), the rate coefficients for Reactions (10.14a) and (10.14b) are 3.5×10^{-15} and 9.3×10^{-12} cm^3 molecules^{-1} s^{-1}, respectively, compared to 6.8×10^{-16} cm^3 molecules^{-1} s^{-1} for k_d in Reaction (10.4). However, whether or not Reactions

(10.14) destroy O_3 faster that Reaction (10.4) depends on the concentrations of NO_2 and O_3. This is illustrated in the following exercise.

Exercise 10.3. If Reaction (10.14b) is the rate-determining step in the Reaction cycle (10.14) derive an expression, in terms of the rate coefficients for Reactions (10.14b) and (10.4) and the concentration of O_3, for the concentration that NO_2 must exceed if the Reaction cycle (10.14) is to destroy O_3 faster than Reaction (10.4).

Solution. Let k be the rate coefficient for Reaction (10.14b). Since this is the rate-determining step, the net Reaction (10.15) cannot proceed faster than the rate at which Reaction (10.14b) destroys atomic oxygen. This rate is given by

$$-\frac{d[O]}{dt} = k[NO_2][O]$$

The rate at which atomic oxygen is destroyed in Reaction (10.4) is

$$-\frac{d[O]}{dt} = k_d[O][O_3]$$

Therefore, the condition for atomic oxygen (and therefore O_3) to be destroyed faster by the Reaction cycle (10.14) than by Reaction (10.4) is

$$k[NO_2][O] > k_d[O][O_3]$$

or

$$[NO_2] > \frac{k_d}{k}[O_3]$$

When the appropriate rate coefficients and concentrations for the various reactions and species in the stratosphere are taken into account, it appears that catalytic cycles of the general form of Reactions (10.12), with X = H, OH, NO, and Cl, all make major contributions to the destruction of O_3 in the stratosphere. Reaction cycle (10.14) dominates in the lower stratosphere, the cycles involving H and OH dominate in the upper stratosphere, and the cycle involving Cl is important in the middle stratosphere. Other halogen and mixed hydrogen-halogen cycles are important in the lower stratosphere (e.g., $ClO + BrO \rightarrow BrCl + O_2$; $HO_2 + ClO \rightarrow HOCl + O_2$; and, $HO_2 + BrO \rightarrow HOBr + O_2$). However, the destruction of O_3 by the various catalytic cycles is not simply additive, because the species in one cycle can react with those in another cycle.

Advanced numerical models of the stratosphere that use the chemical reaction schemes outlined earlier do fairly well in reproducing both the shapes and magnitudes of the measured vertical profiles of O_3 in the natural stratosphere.

10.2 Anthropogenic perturbations to stratospheric ozone[8]

If the concentrations of catalyst X in Reactions (10.12) are increased significantly by anthropogenic activities, the balance between the sources and sinks of atmospheric O_3 will be disturbed and stratospheric O_3 concentrations can be expected to decrease. One of the first concerns in this respect was a proposal in the 1970s to create a fleet of supersonic aircraft flying in the stratosphere. This is because aircraft engines emit nitric oxide (NO), which can decrease odd oxygen by the Reactions (10.14) leading to the net Reaction (10.15). However, this proposal was rejected on both environmental and economic grounds. At the present time, there are not sufficient aircraft flying in the stratosphere to significantly perturb stratospheric O_3.

Of much greater concern, with documented impact, is the catalytic action of chlorine, from human-made chlorofluorocarbons (CFCs), in depleting stratospheric ozone.[9] CFCs are compounds containing Cl, F, and C; CFC-11 ($CFCl_3$) and CFC-12 (CF_2Cl_2) are the most common.[10] CFCs were first synthesized in 1930, as the result of a search for a non-toxic, nonflammable refrigerant. Over the next half-century they became widely used, not only as refrigerants but as propellants in aerosol cans, foaming agents in plastic foam, and as solvents and cleansing agents. Concern about their effects on the atmosphere began in 1973 when it was found that CFCs were spreading globally and, because of their inertness, were expected to have residence times of up to several hundred years in the troposphere.

Such long-lived compounds eventually find their way into the stratosphere, where they absorb UV radiation in the wavelength interval 0.19 to 0.22 μm and photodissociate

$$CFCl_3 + h\nu \rightarrow CFCl_2 + Cl \tag{10.16}$$

and

$$CF_2Cl_2 + h\nu \rightarrow CF_2Cl + Cl \tag{10.17}$$

The chlorine atom released by these reactions can serve as the catalyst X in Reactions (10.12) and destroy odd oxygen in the cycle

$$Cl + O_3 \rightarrow ClO + O_2 \qquad (10.18a)$$

$$ClO + O \rightarrow Cl + O_2 \qquad (10.18b)$$

$$\text{Net:} \quad O_3 + O \rightarrow 2O_2 \qquad (10.19)$$

The first evidence of a significant depletion in stratospheric O_3 produced by anthropogenic chemicals in the stratosphere came, surprisingly, from measurements over the Antarctic. In 1985 British scientists, who had been making ground-based, remote-sensing measurements of O_3 at Halley Bay (76°S) in the Antarctic for many years, reported that there had been about a 30% decrease in total O_3 column abundance each October (i.e., in the austral spring) since 1977. These observations were subsequently confirmed by remote sensing measurements from satellite and by airborne measurements. The satellite measurements show that the region of depleted O_3 over the Antarctic in spring has grown progressively deeper since 1979, and from 1988 to 1997 it occupied an area larger than the Antarctic continent (Figs. 10.3 and 10.4).

The presence of the so-called ozone hole over the Antarctic raised several intriguing questions. Why over the Antarctic? Why during spring? Also, the magnitude of the measured decreases in O_3 were much greater that any predictions based solely on gas-phase chemistry of the type outlined earlier. Why? The answers to these questions provide an excellent demonstration of the maxim that in the atmosphere processes rarely, if ever, act in isolation.

During the austral winter (June–September) stratospheric air over the Antarctic continent is restricted from interacting with air from lower latitudes by a large-scale vortex circulation, which is bound at its perimeter by strongly circulating winds. Very cold air sinks slowly through the center of this vortex (Fig. 10.5). High-level clouds, called *polar stratospheric clouds* (PSCs), form in the cold core of the vortex, where temperatures can fall below –80°C.[11] In the austral spring, as temperature rise, the winds around the vortex weaken, and by November the vortex disappears. However, during winter, the vortex serves as a giant and relatively isolated chemical reactor in which unique chemistry can occur. For example, although the concentrations of O_3 in the vortex are normal in August, the concentrations of ClO in the vortex are ten times greater that just outside the "wall" of the vortex and, by September, O_3 concentrations within the vortex decrease dramatically. There are also sharp decreases in the oxides of nitrogen and water vapor when passing from the outside to the inside of the wall of the vortex. The *denitrification* and

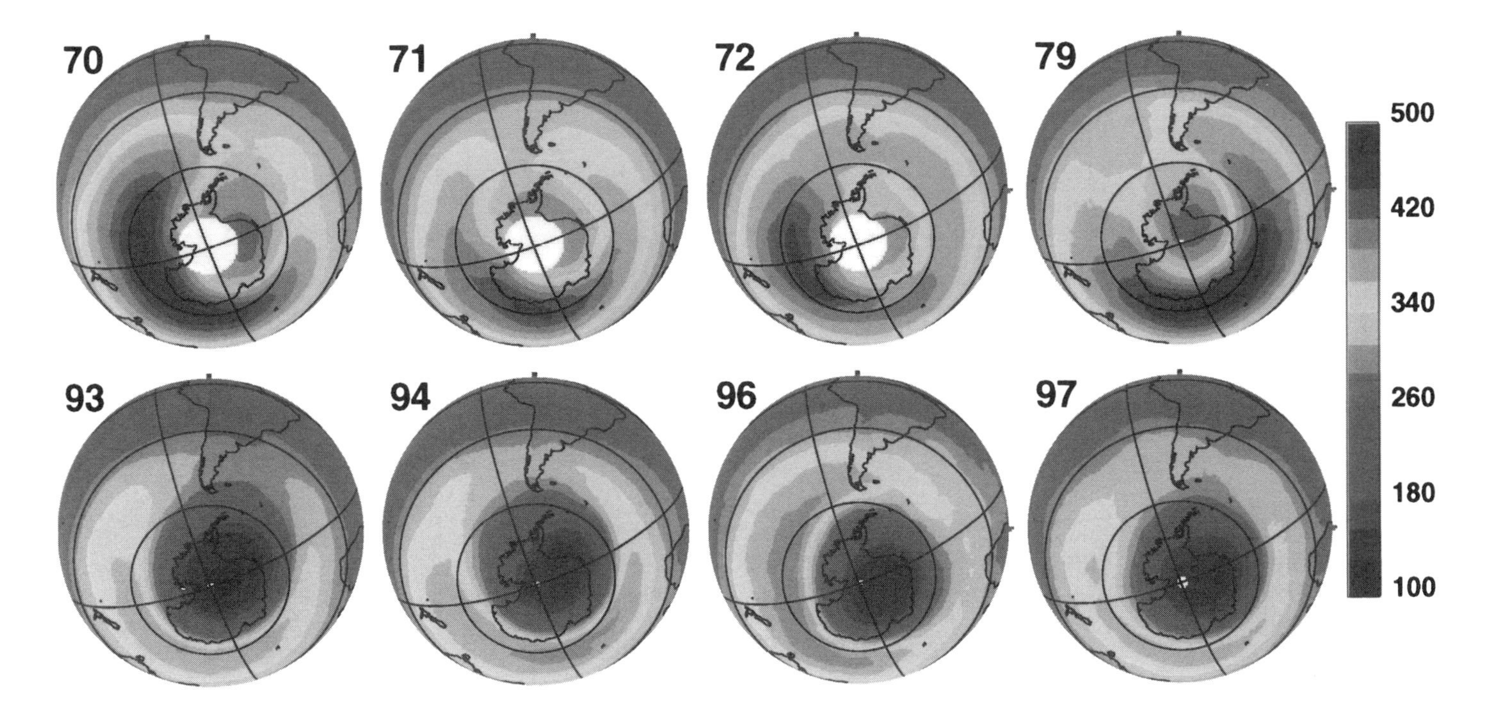

Figure 10.3. Satellite observations of the Antarctic ozone hole in the southern hemisphere during October for the years 1970–1997. The color scale is in Dobson units. (Courtesy P. Newman, NASA Goddard Space Flight Center.) See color section found between page 118 and 119 for a color version of this figure.

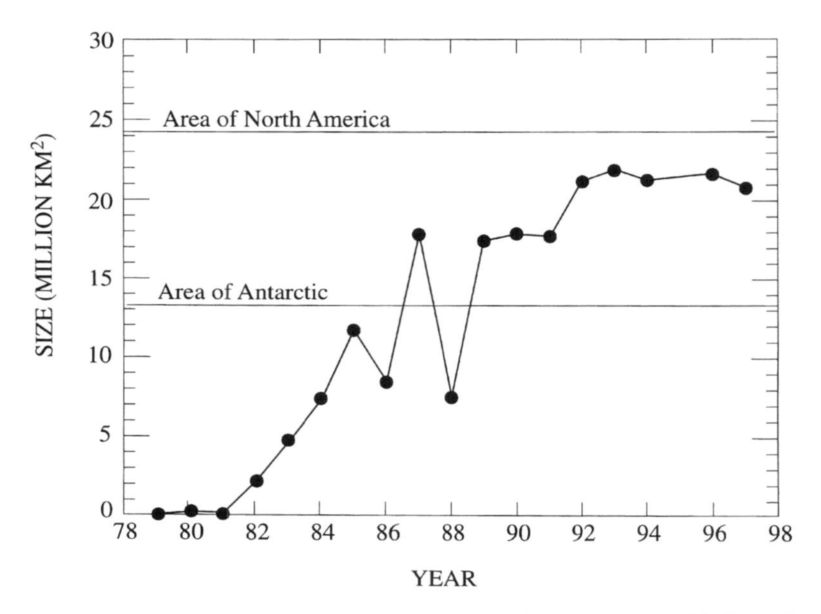

Figure 10.4. Average areal extent of ozone hole (less than 220 DU) from September 7–October 13, 1979–1997. (Courtesy P. Newman, NASA Goddard Space Flight Center.)

dehydration are due, respectively, to the conversion of NO_y to nitric acid (HNO_3), and to the condensation of water at the very low temperatures inside the vortex. These two condensates form three types of PSC. *Type I* PSCs consist of nitric acid trihydrate (NAT) particles, about 1 μm in diameter, which condense at about –80°C. *Type II* PSCs consist of ice-water particles (with nitric acid dissolved in them) about 10 μm in diameter, which form near –85°C. *Type III* PSCs are nacreous ("mother-of-pearl") clouds, due to the rapid freezing of condensed water during flow over topography; Type III PSCs are of limited extent and duration and do not form over the South Pole. As the particles in PSCs slowly sink, they remove both water and nitrogen compounds from the stratosphere. As we shall see, these processes play important roles in depleting O_3 concentrations in the Antarctic vortex.

Most of the Cl and ClO released into the stratosphere (at all latitudes) by Reactions (10.16)–(10.18) are quickly tied up in reservoirs as HCl and $ClONO_2$ by the reactions

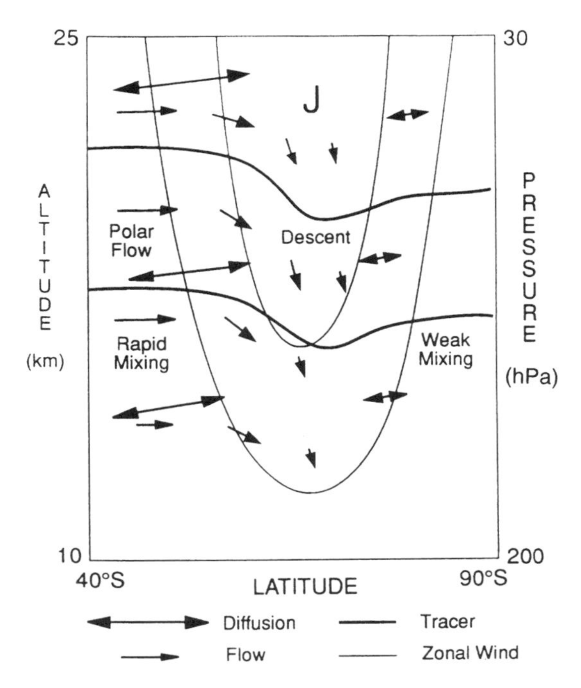

Figure 10.5. Schematic diagram of the circulation and mixing associated with the polar vortex. Thin lines show the zonal contours of winds; J indicates the jet core. Double arrows indicate mixing, with the longer arrows representing larger mixing rates. Single arrows indicate flow directions, with the lengths approximately indicating magnitude. Maximum poleward flow (~0.1 m s^{-1}) occurs equatorward of the wind jet. The largest descent zone (with vertical velocities ~0.05 cm s^{-1}) is roughly coincident with the jet core. Long-lived tracer isopleths are shown by thick lines. This picture of the polar vortex is for the Arctic midwinter period or the Antarctic early spring. [From *Scientific Assessment of Ozone Depletion: 1991*, World Meteorological Organization, p. 4.11 (1992).]

$$Cl + CH_4 \rightarrow HCl + CH_3 \tag{10.20}$$

$$ClO + NO_2 + M \rightarrow ClONO_2 + M \tag{10.21}$$

Liberation of the active Cl atom from these reservoirs is generally slow. However, on the surface of the ice particles that form PSCs, the following catalytic (catalyzed by the ice particle) reaction can occur

$$ClONO_2(s) + HCl(s) \rightarrow Cl_2(g) + HNO_3(s) \tag{10.22}$$

where the parenthetical *s* has been inserted to emphasize those compounds that are on (or in) ice particles. The nitric acid remains with the ice particles, but the Cl_2 is released as a gas that is photodissociated in the stratosphere

$$Cl_2 + h\nu \rightarrow Cl + Cl \tag{10.23}$$

In addition to catalyzing Reaction (10.22), the ice particles play another role: they can fall out and remove nitrogen from the stratosphere (as HNO_3), which reduces the $ClONO_2$ (s) reservoir that ties up Cl and ClO. Thus, on both counts, during the austral winter the ice particles that comprise PSCs in the Antarctic vortex set the stage for the destruction of ozone by enhancing the concentrations of active ClO and Cl. However, Reactions (10.18) cannot proceed with full vigor until enough sunlight is present to release both sufficient free Cl atoms (by Reaction 10.23) *and* sufficient quantities of atomic oxygen (by Reaction 10.3). Since there is not enough sunlight for this purpose in early spring in the Antarctic stratosphere, Reactions (10.18) alone cannot explain the very large depletions of O_3 that produce the Antarctic O_3 hole, although they probably contribute to it.

A cycle catalyzed by ClO that appears capable of explaining about three-quarters of the observed O_3 loss in the Antarctic ozone hole is[12]

$$ClO + ClO + M \rightarrow (ClO)_2 + M \tag{10.24a}$$

$$(ClO)_2 + h\nu \rightarrow Cl + ClOO \tag{10.24b}$$

$$ClOO + M \rightarrow Cl + O_2 + M \tag{10.24c}$$

$$2Cl + 2O_3 \rightarrow 2ClO + 2O_2 \tag{10.24d}$$

$$\text{Net:} \quad 2O_3 + h\nu \rightarrow 3O_2 \tag{10.25}$$

The following points should be noted about this reaction cycle.

- Reactions (10.24) form a catalytic cycle in which ClO is the catalyst, because two ClO molecules are regenerated for every two ClO molecules that are consumed.
- The cycle does not depend on atomic oxygen (which is in short supply).
- The Cl atom in the ClO on the left side of (10.24a) derives from Cl released from CFC via Reactions (10.16) and (10.17). However, as we have seen, the Cl atom is then normally quickly tied up as HCl and $ClONO_2$ by Reactions (10.20) and (10.21). But in the presence of PSCs, Cl_2 gas is released by Reaction (10.22) and, as soon as the solar radiation reaches sufficient intensity in early spring, Reaction (10.23)

releases Cl. Reaction (10.24d) converts this into ClO, which is then available for the first step in the Reaction cycle (10.24) that leads to the rapid depletion of O_3 in the Antarctic stratosphere.

- The dimer $(ClO)_2$ is formed by Reaction (10.24a) only at low temperatures. Low enough temperatures are present in the Antarctic stratosphere, where there are also large concentrations of ClO. Therefore, on both counts, the Antarctic stratosphere in spring is a region in which the Reaction cycle (10.24) can destroy large quantities of O_3.

Exercise 10.4. If Reaction (10.24a) is the slowest step in the catalytic cycle (10.24), and the pseudo first-order rate coefficient for this reaction is k, derive an expression for the amount of O_3 destroyed over a time period ΔT by this cycle.

Solution. Inspection of the Reaction cycle (10.24) shows that two O_3 molecules are destroyed in each cycle. Also, the rate of the cycle is determined by the slowest reaction in the cycle, namely Reaction (10.24a). Therefore, the rate of destruction of O_3 by this cycle is

$$\frac{d[O_3]}{dt} = -2k[ClO]^2$$

(where [M] has been incorporated into the pseudo first-order rate coefficient k).

Hence, over a time period Δt, the amount of ozone destroyed, $\Delta[O_3]$, is

$$\int_0^{\Delta[O_3]} d[O_3] = -2k\int_0^{\Delta t} [ClO]^2\, dt$$

or, if [ClO] does not change in the period Δt,

$$\Delta[O_3] = -2k[ClO]^2\, \Delta t$$

At this point, the reader might well ask whether an ozone hole develops in the Arctic stratosphere in winter and, if not, why not? In the Arctic, depletions of 15% are thought to be typical, and 30% losses were observed in the 1992–1993 winter. There is evidence that these losses are due to anomalous chlorine chemistry similar to that which occurs in the Antarctic. For example, in a field study carried out in the Arctic in 1988–1989, sharp increases were measured in the concentrations of ClO in the stratosphere, and these appeared to be associated with PSCs. Also, increases in OClO were measured, which provides some support for the

Reaction cycle (10.24).[13] On the other hand, although some denitrification was measured at altitudes around 20 km, it was not as great as in the Antarctic stratosphere, perhaps because the PSCs evaporated in the lower stratosphere. Also, dehydration was much less in the Arctic. In any case, the decrease in the total O_3 column in the Arctic in 1988–1989 was only a few percent, much less that observed in the Antarctic. It is not known whether this was due to insufficient "anomalous" chemistry of the type that produces the Antarctic ozone hole or to meteorological conditions that were less suitable for O_3 depletion. For example, stratospheric temperatures remained very low until the middle of February in 1989, when there was a sudden warming and the PSC disappeared. Thus, air that was sufficiently cold for Reaction (10.24a) to proceed rapidly may not have received sufficient solar radiation for Reaction (10.24b) to proceed effectively. In the Antarctic, stratospheric O_3 is depleted primarily in September (which corresponds to March in the Arctic) when temperatures are still very low, but solar radiation is increasing rapidly. It would appear that while concentrations of CFC remain high in the atmosphere, the Arctic stratosphere has the potential to cause the same dramatic losses in O_3 as the Antarctic stratosphere. However, the combination of chemical and meteorological conditions that lead to such reductions may not be as common in the Arctic as in the Antarctic.

On a global scale, ground-based and satellite measurements show decreases in total column O_3 at midlatitudes in the northern hemisphere of 2.7% per decade in winter, 1.3% per decade in summer, and 1.2% per decade in the fall. For example, in northern mid- and polar latitudes, O_3 levels in the last two months of 1996 were 5% to 8% below the 1957–1970 averages. Similar decreases are apparent at midlatitudes in the Southern Hemisphere; at high latitudes, beneath the region of the Arctic ozone hole, the decreases are 14% per year. The decreases have occurred primarily in the lower stratosphere. No trends in O_3 concentrations have been observed in the tropics. The chemical and dynamical mechanisms responsible for these losses are not, at present, understood.

Concerns about the health and environmental hazards of increased UV radiation at the Earth's surface, which accompany depletion in the total column O_3, led to international agreements to reduce the manufacture of CFCs, and to eliminate them completely by the year 2000. Consequently, CFC amounts in the lower atmosphere are no longer increasing, and their *rate* of growth in the stratosphere is decreasing. However, due to the long lifetimes of CFCs, the concentrations of Cl in the stratosphere are expected to continue to rise for some time. There-

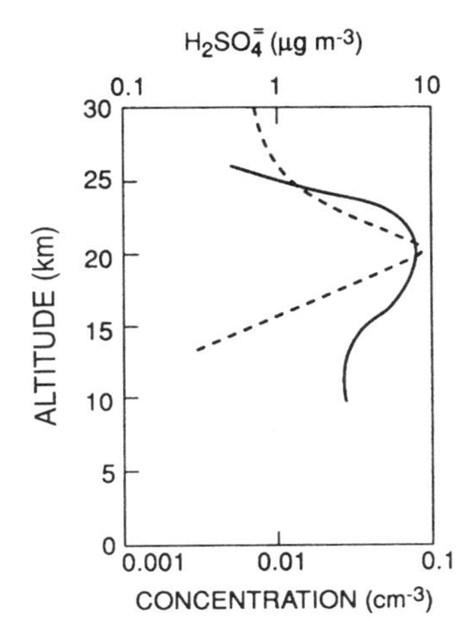

Figure 10.6. Vertical profiles in the lower stratosphere of particles with radius ~0.1 to 2 μm (solid line and bottom scale) and mass concentration of liquid sulfuric acid at standard temperature and pressure (dashed line and upper scale).

fore, further decreases in total column O_3, and over increasingly larger regions of the globe, are to be expected.[14]

10.3 Stratospheric aerosols; sulfur in the stratosphere

Aitken nucleus concentrations show considerable variations in the lower stratosphere, although they generally decrease slowly with height. In contrast, aerosols with radii ~0.1 to 2 μm reach a maximum concentration of ~0.1 cm^{-3} at altitudes of ~17 to 20 km (Fig. 10.6). Since these aerosols are composed of about 75% sulfuric acid (H_2SO_4) and ~25% water, the region of maximum sulfate loading in the lower stratosphere is called the *stratospheric sulfate layer*, or sometimes the *Junge layer*, after C. Junge who discovered it in the late 1950s.

Stratospheric sulfate aerosols are produced primarily by the oxidation of SO_2 to H_2SO_4 vapor in the stratosphere

$$SO_2 + OH + M \rightarrow HOSO_2 + M \tag{10.26a}$$

then,

$$HOSO_2 + O_2 \rightarrow HO_2 + SO_3 \tag{10.26b}$$

Also,

$$SO_2 + O + M \rightarrow SO_3 + M \qquad (10.27a)$$

then,

$$SO_3 + H_2O \rightarrow H_2SO_4 \qquad (10.27b)$$

The conversion of the H_2SO_4 from vapor to liquid can occur by two main mechanisms:

- The combination of molecules of H_2SO_4 and H_2O (i.e., *homogeneous, bimolecular nucleation* – see Section 7.8), and/or the combination of H_2SO_4, H_2O, and HNO_3 to form new (primarily sulfuric acid) droplets (referred to as *homogeneous, heteromolecular nucleation*).
- Vapor condensation of H_2SO_4, H_2O, and HNO_3 onto the surfaces of preexisting particles with radius $>0.15\,\mu$m (this is referred to as *heterogeneous, heteromolecular nucleation*).

Model calculations suggest that the second mechanism is the more likely route in the stratosphere. The tropical stratosphere is probably the major region where the nucleation process occurs, and the aerosols are then transported to higher latitudes by large-scale atmospheric motions. The most significant source of the SO_2 are major volcanic eruptions that can inject large quantities of SO_2 into the stratosphere, which converts to H_2SO_4 with an *e*-folding time of about 1 month. The effect of such eruptions on the stratospheric aerosol optical depth (which is a measure of the aerosol loading) is shown in Figure 10.7. The 1978–1979 period is generally referred to as the stratospheric aerosol "background" state, because it followed a period of about five years without volcanic eruptions. The 1982 El Chichon volcanic eruption produced the largest perturbation to the stratospheric sulfate layer observed during the 1980s, and the eruption of Mount Pinatubo in June 1991, which appears to have been the largest volcanic eruption of the twentieth century, had an even larger effect on stratospheric aerosols. The enhancements in aerosol loadings in the Antarctic during the local winter and early spring, which can be seen in Figure 10.7, are due to the PSC formation discussed in Section 10.2. In fact, the NAT particles, which are the major component of the Type I PSCs, condense onto the particles in the stratospheric sulfate layer. The results shown in Figure 10.7 were obtained from a satellite by measuring the attenuations and scattering of solar radiation as it passed tangentially through the Earth's atmosphere. This capability is now providing virtually continuous global monitoring of stratospheric aerosols and many trace gases.

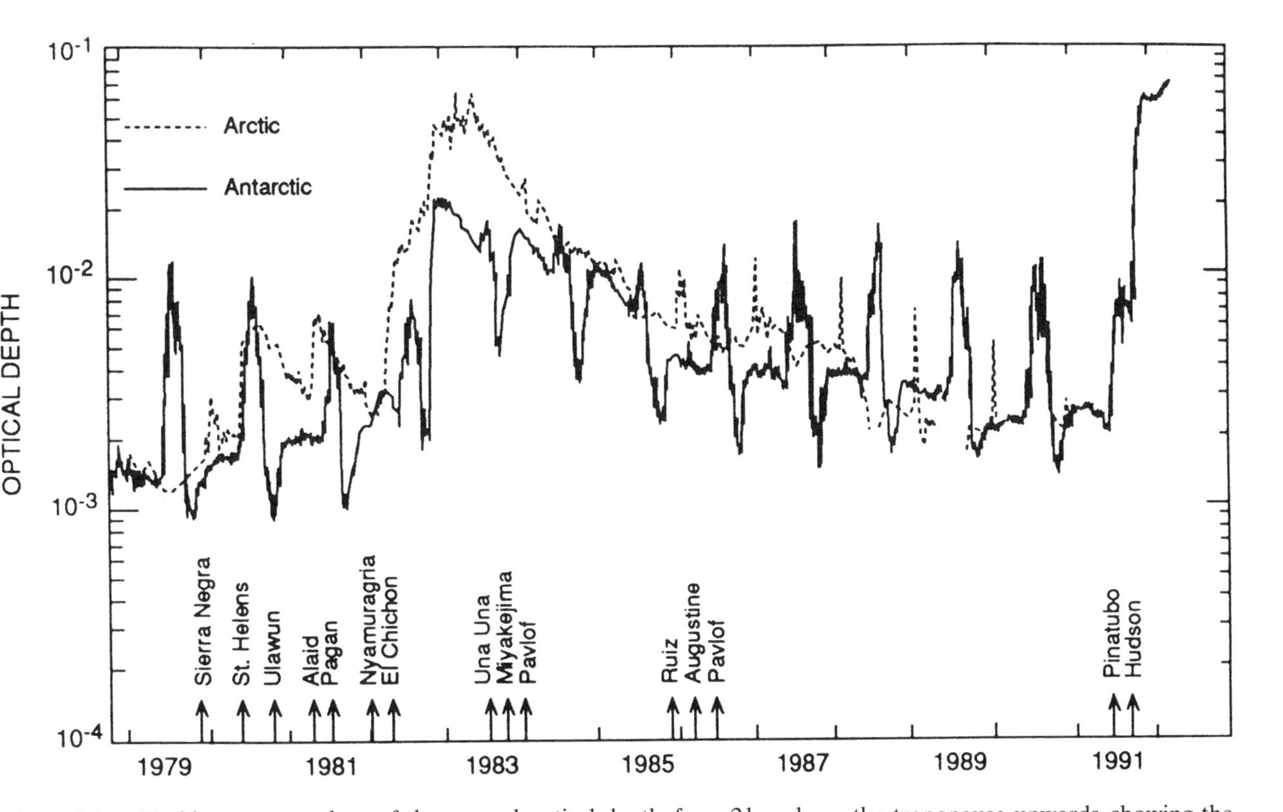

Figure 10.7. Weekly average values of the aerosol optical depth, from 2 km above the tropopause upwards, showing the effects of the large volcanic eruptions of El Chichón in 1982 and Mount Pinatubo in June 1991. The periodic increases in optical depth in the Antarctic in the austral winter and spring are due to polar stratospheric clouds. [From M. P. McCormick in *Aerosol-Cloud-Climate Interactions*, Ed. P. V. Hobbs, Academic Press, p. 210 (1993).]

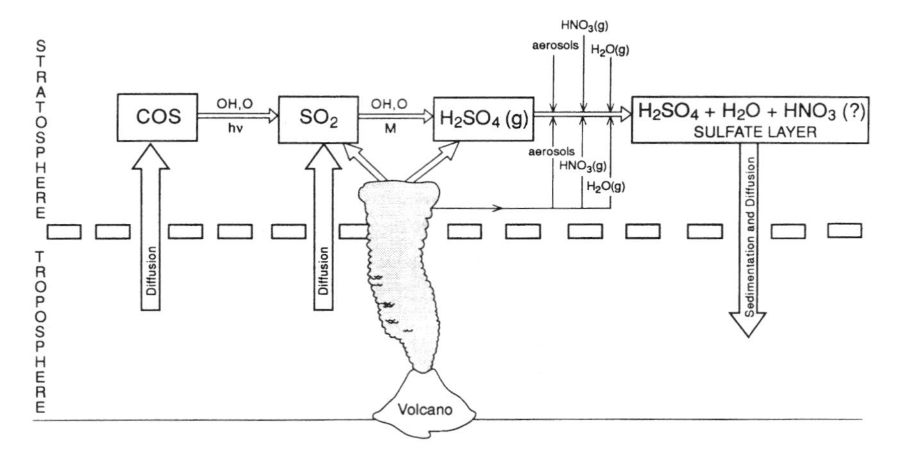

Figure 10.8. Schematic diagram of the processes responsible for the stratospheric sulfate layer.

Enhancement of the sulfate layer by volcanic eruptions can cause depletions in stratospheric O_3 due to the H_2SO_4 droplets acting to modify the distribution of catalytically active free radicals. For example, there was dramatic O_3 depletion in the stratosphere following the eruption of Mount Pinatubo. The reflection of shortwave solar radiation and the absorption of longwave terrestrial radiation by volcanic aerosol is well established. Thus, satellite measurements revealed a 1.4% increase in solar radiation reflected from the atmosphere for several months after the eruption of Mount Pinatubo. Also, for two months after this eruption there was a ~3°C increase in the daily zonal mean stratospheric temperature at low latitudes. Since the particles produced by volcanic eruptions drift slowly toward the Earth's surface, their effects generally last for just a year or so. However, because the eruption of Mount Pinatubo caused such a major perturbation, the chemical effects continued for at least three years.

When major volcanic activity is low, the primary source of gaseous sulfur compounds that maintain the stratospheric sulfate layer is believed to be the transport of carbonyl sulfide (COS) and SO_2 across the tropopause (Fig. 10.8). COS can be converted to SO_2 as follows

$$COS + h\nu \rightarrow CO + S \tag{10.28a}$$

$$S + O_2 \rightarrow SO + O \tag{10.28b}$$

$$O + COS \rightarrow SO + CO \quad (10.28c)$$

$$SO + O_2 \rightarrow SO_2 + O \quad (10.28d)$$

$$SO + O_3 \rightarrow SO_2 + O_2 \quad (10.28e)$$

also,

$$COS + OH \rightarrow \text{products} \rightarrow SO_2 \quad (10.29)$$

In addition to converting COS to SO_2, the Reactions (10.28) both destroy and produce odd oxygen, but more odd oxygen is probably formed by this mechanism than is destroyed. Based on the observed response of the sulfate layer to volcanic eruptions, the conversion of SO_2 to H_2SO_4 vapor, and then to H_2SO_4 droplets, takes about a month.

The mixing ratio of COS decreases with height in the lower stratosphere (from about 0.4 ppbv at the tropopause to 0.02 ppbv at 30 km), the concentration of SO_2 remains roughly constant (at ~0.05 ppbv), and the concentration of liquid H_2SO_4 peaks at ~20 km. This supports the idea that COS is converted to SO_2, which then forms H_2SO_4 condensate by the mechanisms discussed previously. Numerical modeling results indicate that the direct transfer of SO_2 into the stratosphere from the troposphere is also important. The model calculations also show that H_2SO_4 (and O_3) are produced at low latitudes in the stratosphere, with maximum transport toward the poles in winter and spring.

Exercises

See Exercises 1(dd)–(mm) and Exercises 47–54 in Appendix I.

Notes

1 Electromagnetic radiation in this wavelength range is dangerous to living cells. Radiation with a wavelength (λ) $\leq 0.29\,\mu$m is lethal to lower organisms and to the cells of higher organisms. Radiation with λ from $0.29\,\mu$m to $0.32\,\mu$m (*UV-B radiation*) has adverse effects on human health and on animals and plants. Were it not for O_3 in the stratosphere, radiation from the Sun with λ from $0.23\,\mu$m to $0.32\,\mu$m, would reach the Earth's surface unhindered. Ozone strongly absorbs UV radiation in just this wavelength band. For this reason, less than 1 part in 10^{30} of the flux of solar radiation at the top of the atmosphere with $\lambda = 0.25\,\mu$m reaches the Earth's surface. As we have seen in Chapter 1, the absorption of UV radiation by O_3 was essential for the emergence of life on Earth.

2 It is common to use nanometers (nm; $1\,\mu$m $= 10^3$ nm) as the unit of wavelength in the ultraviolet and visible regions. However, for consistency with earlier chapters, we will use micrometers.

3 W. N. Hartley (1846–1913). Spectroscopist. Professor of chemistry at Royal College of Science, Dublin.

4 G. M. B. Dobson (1889–1976). English physicist and meteorologist. Obtained the first measurement of the variation of wind with height using pilot balloons (1913). In 1922 Dobson discovered (with F. A. Lindemann) the presence of a warm layer of air at ~50 km, which he attributed to the absorption of UV radiation by O_3. Dobson built a UV solar spectrograph for measuring the atmospheric O_3 column. He also obtained the first measurements of water vapor in the stratosphere.

5 Sydney Chapman (1888–1970). English geophysicist. Chapman made important contributions to a wide range of geophysical problems, including geomagnetism, space physics, photochemistry, and diffusion and convection in the atmosphere.

6 "Odd oxygen" refers to [O] + [O_3]. In the Chapman reactions odd oxygen is produced only by Reaction (10.1) and lost only in Reaction (10.4). Reactions (10.2) and (10.3) interconvert atomic oxygen and O_3 and determine the ratio [O]/[O_3]; both reactions are fast during the day. Therefore, atomic oxygen and O_3 are interconverted rapidly. Below about 45 km in the stratosphere, O_3 accounts for >99% of odd oxygen.

7 Of course, the Chapman reactions would not continue indefinitely, since a new photostationary state would ultimately be reached.

8 Section 10.2 is based, in part, on a very readable account of perturbations to stratospheric O_3 in *Chemistry of Atmospheres* by R. P. Wayne, Oxford University Press, Oxford (1991).

9 In 1995, Paul Crutzen, Mario Molina, and Sherwood Rowland were awarded the Nobel Prize in chemistry for predicting stratospheric O_3 loss by CFCs and nitrogen-containing gases.

10 Compounds containing Cl, F, C, and H are called *hydrochlorofluorocarbons* (HCFC); they are CFC replacements.

11 During the drift of the *Deutschland* in the Weddell Sea in 1912, polar stratospheric clouds (PSCs) were observed in the late winter and early spring. Also, the Norwegian-British-Swedish 1949–1952 expedition to the Antarctic frequently observed a thin "cloud-veil" in the lower stratosphere in winter months. However, the widespread nature and periodicity of PSC formation in the Antarctic was not fully appreciated until they were observed by satellites starting in 1979.

12 A reaction cycle involving ClO and BrO, and another involving Cl and H radicals, together appear more than capable of explaining the remainder of the ozone loss.

13 The symmetric form of chlorine dioxide (OClO) is different from the unstable species ClOO in Reaction (10.24b). However, the presence of OClO provides an important indication of the amount of ClO, and therefore the destruction of odd oxygen, in the O_3 hole.

14 The O_3 hole over Antarctica in 1998 was almost as severe as any seen before. It extended over an area of about 26 million square kilometers (larger than North America), and O_3 was totally absent from altitudes between 15 and 21 km. In early October 1998, the total O_3 column over the South Pole was only 92 DU. Prior to 1998, the O_3 hole was deeper than this only in 1993, when the catalytic effect of particles from the 1991 eruption of Mt. Pinatubo drove the Antarctic O_3 column down to 88 DU.

The large depletion of O_3 in 1998 was likely due to the unusually low stratospheric temperatures in that year, which increased the occurrence of PSCs that catalytically enhance the destruction of O_3 by CFCs. It has been postulated (based on numerical models) that lower temperatures in the stratosphere will become increasingly common as greenhouse gases accumulate in the atmosphere. This is because, while greenhouse gases warm the lower atmosphere, they cool the stratosphere by radiating heat to space.

Appendix I

Exercises[a]

1. Explain, answer or interpret the following:
   (a) Although nitrogen and oxygen form compounds with each other, they occur largely uncombined in the Earth's atmosphere.
   (b) The percentage of the total oxygen in the atmosphere that is in the form of atomic oxygen increases rapidly with increasing altitudes above ~80 km, and above ~180 km all of the oxygen is atomic oxygen.
   (c) Oceans do not exist on Mars or Venus.
   (d) Oxygen is virtually absent in the atmospheres of Mars and Venus.
   (e) Nitrogen is the dominant constituent in the Earth's atmosphere, but it is only a minor constituent in the atmospheres of Mars and Venus.
   (f) The carbon-14 content of atmospheric CH_4 samples (collected before contamination of the atmosphere with carbon-14 from nuclear explosions) suggested that ~80% of the CH_4 was from the decay of recent organic materials. Hence, it can be deduced that <20% of the CH_4 in the atmosphere derives from old fossil sources (such as natural gas leakage).
   (g) On a global scale the concentration of CO_2 is the same

[a] Exercises marked by an asterisk are more difficult. Answers to all of the exercises and solutions to the more difficult exercises are given in Appendix II. If you have difficulty understanding the basic chemical aspects of any of the exercises, it is recommended that you study *Basic Physical Chemistry for the Atmospheric Sciences* by P. V. Hobbs, Cambridge University Press, Cambridge (2000).

all over the world, whereas the concentration of hydrogen sulfide varies considerably from one location to another.

(h) What is the major hydrocarbon in air?

(i) The detection of radon over the oceans is a good indicator of the recent intrusion of continental air.

(j) Why do objects viewed in direct sunlight, particularly around sunrise and sunset have a reddish color?

(k) The contribution of scattering by molecules to the attenuation of the solar irradiance diminishes with increasing wavelength (see Fig. 4.1).

(l) How would you expect the heating rates shown in Figure 4.3 to change with solar zenith angle (θ)?

(m) With reference to Figure 4.11, write down the energy balance for the surface of the Earth. Hence show that the surface temperature of the Earth is higher than the emission temperature (T_E), defined by Eq. (4.41), because the atmosphere is transparent to solar radiation but augments solar heating of the surface by its own downward emission of longwave radiation.

(n) Show that the longwave radiation received by the Earth's surface due to emission from the atmosphere is equal to the solar heating of the surface.

(o) The major anion in ground water beneath limestone formations is the bicarbonate ion (HCO_3^-).

(p) In the shadow of a thick cloud the concentration of the hydroxyl radical falls to nearly zero.

(q) About how many hydroxyl radicals are there in $1\,cm^3$ of air at midday?

(r) Which three trace constituents primarily determine the oxidizing capacity of the atmosphere during daytime?

(s) What sulfur compound is the major reservoir of sulfur in the troposphere?

(t) Use Figure 6.1 to estimate the range of scale heights for aerosol number concentrations near the Earth's surface in remote continental air.

(u) Use Figure 6.2 to estimate the approximate scale height for aerosol mass concentrations for remote continental air. Why is this value near the lower end of the range for the scale heights for aerosol number concentrations (see Exercise (t))?

(v) Theoretical predictions indicate that thermophoresis should dominate diffusiophoresis for aerosols with diameters <1 μm. Observation shows a dust-free space around evaporating drops. Do these observations confirm the theoretical predictions?

(w) Outline a hypothetical scenario in which the effects on clouds of DMS emissions from the ocean, which might possibly be increased by global warming, could act as a "thermostat" that would tend to offset global warming.

(x) The residence time of water vapor in middle latitudes is ~5 days but in the polar regions it is ~12 days.

(y) Since the main sink for CH_4 is oxidation by the OH radical, one might expect that CH_4 concentrations in high latitudes would be lower in summer (when OH concentrations are high) than they are in winter. However, measurements of CH_4 concentrations do not show large seasonal trends. Suggest a possible explanation. (*Hint*: what is the major source of CH_4 in high latitudes?)

(z) The most important pathways of sulfur through the troposphere involve injection as H_2S, DMS, COS, and CS_2 (see Fig. 8.3). What are the oxidation states of sulfur in these gases? An important sink for sulfur gases is conversion to sulfate ($SO_4^=$). What is the oxidation state of sulfur in $SO_4^=$?

(aa) Even the cleanest combustion fuels (e.g., hydrogen) are sources of NO_x.

(bb) Tall chimneys (or stacks) on industrial plants might replace one air pollution problem by another.

(cc) To what extent do the time variations of NO, NO_2, and O_3 shown in Figure 9.1 agree and disagree with the qualitative predictions of Eq. (9.6)?

(dd) Give a qualitative explanation why the Chapman reactions (see Section 10.1) predict a peak concentration in ozone at some level in the atmosphere.

(ee) The Chapman reactions predict that the ratio of $[O]/[O_3]$ increases with increasing altitude.

(ff) Long-range transport of chemical species plays a more important role in ozone chemistry in the lower stratosphere than in the upper stratosphere.

(gg) In the lower stratosphere, the concentrations of atomic oxygen decrease rapidly when the sun sets (*Hint*: Consider the Chapman reactions).

(hh) Regular diurnal variations in stratospheric ozone are much larger at high altitudes ($\gtrsim$40 km) than at lower altitudes (*Hint*: Consider the Chapman reactions).

(ii) The Chapman reactions produce no net change in concentration of "odd" oxygen.

(jj) Why are catalytic cycles of the form of Reactions (10.12), rather than noncatalytic reactions involving trace chemical species, required to explain the removal of stratospheric ozone?

(kk) For catalytic cycles such as Reactions (10.12) to be efficient, each reaction comprising the cycle must be exothermic.

(ll) The tropospheric contributions of CS_2, H_2S, and DMS to the stratospheric sulfur budget are probably small.

(mm) In the stratosphere the maximum amounts of ozone, expressed as a partial pressure, occur at lower altitudes than when they are expressed in ppmv (see Fig. 10.2).

2. Water can be photolyzed by the reaction series

$$H_2O + h\nu \rightarrow OH + H$$

$$OH + OH \rightarrow O + H_2O$$

$$O + OH \rightarrow O_2 + H$$

$$O + O + M \rightarrow O_2 + M$$

Balance this *series* of reactions, and write down the net reaction. How many water molecules are needed to produce one molecule of O_2 by these reactions? (Note: the photolysis of H_2O produces a net gain in O_2 only if the atomic hydrogen released by the above series of reactions escapes from the top of the atmosphere. Otherwise, H_2O is reformed.)

3. The combination of CO_2 photolysis with the photolytic reaction series given in Exercise 2 yields

$$H_2O + h\nu \rightarrow OH + H$$

$$CO_2 + h\nu \rightarrow CO + O$$

$$CO + OH \rightarrow CO_2 + H$$

$$OH + OH \rightarrow O + H_2O$$

$$O + O + M \rightarrow O_2 + M$$

What is the net balanced reaction? Does this alter the conclusions arrived at in Exercise 2?

4. It has been suggested that hydrogen in the Earth's primitive atmosphere led to the production of CH_4 by the reaction

$$CO_2(g) + 4H_2(g) \rightleftarrows CH_4(g) + 2H_2O(g)$$

(a) The equilibrium constants for this reaction at 300 and 400 K are 5.2×10^{19} and $2.7 \times 10^{12}\,bar^{-2}$, respectively. If the partial pressures of H_2O, CO_2, and H_2 in the primitive atmosphere are taken to be 3.0×10^{-2}, 3.0×10^{-4}, and 5.0×10^{-5} bar, respectively, what are the equilibrium pressures of CH_4 at 300 and 400 K?

(b) At 400 K the rate coefficient for the above reaction is large, but at 300 K it is immeasurably small. Is it likely that this reaction was responsible for the conversion of much H_2 into CH_4 in the primitive atmosphere?

5. A possible route for CH_4 production in the Earth's primitive atmosphere is the promotion of the reaction given in Exercise 4 by methanogenic bacteria (one of the most ancient forms of life). These bacteria use this reaction as a source of heat, with the hydrogen being derived from the degradation of alcohols by nonmethanogenic bacteria in the same environment.

(a) Do you think this mechanism is more likely to have been responsible for the production of CH_4 in the Earth's primitive atmosphere than that given in Exercise 4? Why?

(b) How much heat would the methanogenic bacteria receive per mole of CH_4 produced by the reaction given in Exercise 4 at 25°C and 1 atm? (The heats of formation of

CO_2(g), H_2(g), CH_4(g), and H_2O(g) at 25°C and 1 atm are –393.5, 0, –74.8, and –241.8 kJ mol^{-1}, respectively.)

*6. If the use of fossil fuels continues to expand as it has done over the past few decades, about 3.0×10^{13} kg of fossil fuels will be consumed annually by the year 2000. If half of the resulting CO_2 remained in the air, what would be the annual rate of increase (in ppmv) in atmospheric CO_2 around the year 2000? Assume that the fuels are 80% carbon (C) by mass. (Mass of the Earth's atmosphere = 5.1×10^{18} kg. Atomic weights of carbon and oxygen are 12 and 16, respectively, and the apparent molecular weight of air is 29.)

7. The rate of decay of a chemical involved in a reaction that is second order (bimolecular) in one reactant A is given by

$$-\frac{d[\mathrm{A}]}{dt} = k[\mathrm{A}]^2$$

where k is a constant. Derive an expression for the half-life of A in terms of k and the concentration of A at time $t = 0$ (A_0).

8. The half-life of a first-order chemical reaction is 20.2 s. How much of a 10.0 g sample of an active reactant in the reaction will be present after 60.6 s?

9. An airshed is sufficiently large that it can be considered to have a constant volume of 6×10^5 m^3. The airshed initially contains 5% by volume of a chemical X. If more of the chemical X enters the airshed at a rate of 500 m^3 per minute, what is the initial (instantaneous) residence time of chemical X in the airshed with respect to (a) the influx of X, and (b) the efflux of X?

10. If the concentration of NH_3 in air is 0.456 μg m^{-3} at 0°C and 1 atm, what is its concentration in ppbv? (Atomic weights of H and N are 1.01 and 14.0, respectively. The total number of molecules in 1 m^3 of air at 1 atm and 0°C is 2.69×10^{25}.)

11. If the concentration of NO_2 in air is 28.125 μg(N) m^{-3} at 0°C and 1 atm, what is its concentration in ppbv? (Atomic weights of N and O are 14 and 16, respectively.)

12. If all of a well-mixed trace gas in the atmosphere (e.g., argon)

were reduced to the density of the gas at ground level (ρ_0), show that the depth of the gas would be $\rho_0 H$, where H is the scale height of the gas.

13. If a woman lives to an age of 78, what percentage of the particular O_2 molecules that were in the atmosphere when she was born will be there when she dies? (Assume a residence time for O_2 in the atmosphere of 5,000 years.)

14. Assuming that the atmosphere consists of a homogeneous layer of air from sea-level to 3 km with an extinction coefficient of $1 \times 10^{-4}\,m^{-1}$, another homogeneous layer from 3 to 10 km with an extinction coefficient of $3 \times 10^{-5}\,m^{-1}$, and a third homogeneous layer from 10 to 20 km with an extinction coefficient of $1 \times 10^{-6}\,m^{-1}$, what is the total column optical depth of the atmosphere?

15. What is the scale height of an isothermal dry atmosphere with a temperature of 15°C? (The gas constant of 1 kg of dry air is $287\,J\,deg^{-1}\,kg^{-1}$, and the acceleration due to gravity is $9.81\,m\,s^{-2}$.) What is the physical interpretation of the scale height you have calculated?

16. The solar constant, $E_{\lambda\infty}$, and the total atmospheric optical depth, τ_λ, can be derived by measuring the irradiance at ground level, E_λ, for various solar zenith angles, θ. How? [*Hint*: Generalize Eq. (4.11).]

17. Calculate the value of the photolysis rate coefficient j for Reaction (4.46) between wavelengths of 0.295 and 0.410 μm and at 20°N latitude for (a) 1 March, and (b) 1 August for the conditions listed in Table A.1. Assume cloud-free conditions and a surface albedo of zero. Use the values for the quantum yields and absorption cross-sections given in Table 4.2. (c) Why is the actinic flux higher on 1 August than 1 March? (d) If the number concentration of NO_2 molecules is $5 \times 10^{15}\,m^{-3}$, what is the instantaneous rate of dissociation of NO_2 molecules by photolysis under the conditions given earlier for 1 March?

Table A.1. *Actinic fluxes from 0.295–0.410 μm wavelength at solar noon, at the surface, and under cloud-free conditions, on 1 March and 1 August at 20°N latitude*

Wavelength Interval (μm)	Actinic Flux on 1 March (photons $cm^{-2}s^{-1}$)	Actinic Flux on 1 August (photons $cm^{-2}s^{-1}$)
0.295–0.300	1.70×10^{12}	3.70×10^{12}
0.300–0.305	2.34×10^{13}	3.62×10^{13}
0.305–0.310	9.65×10^{13}	1.28×10^{14}
0.310–0.315	2.36×10^{14}	2.85×10^{14}
0.315–0.320	3.42×10^{14}	3.94×10^{14}
0.320–0.325	4.38×10^{14}	4.96×10^{14}
0.325–0.330	6.40×10^{14}	7.14×10^{14}
0.330–0.335	6.84×10^{14}	7.54×10^{14}
0.335–0.340	6.81×10^{14}	7.45×10^{14}
0.340–0.345	7.38×10^{14}	8.03×10^{14}
0.345–0.350	7.40×10^{14}	8.02×10^{14}
0.350–0.355	8.42×10^{14}	9.08×10^{14}
0.355–0.360	7.88×10^{14}	8.36×10^{14}
0.360–0.365	8.64×10^{14}	9.25×10^{14}
0.365–0.370	1.07×10^{15}	1.14×10^{15}
0.370–0.375	9.65×10^{14}	1.03×10^{15}
0.375–0.380	1.07×10^{15}	1.15×10^{15}
0.380–0.385	8.95×10^{14}	9.48×10^{14}
0.385–0.390	9.87×10^{14}	1.04×10^{15}
0.390–0.395	1.02×10^{15}	1.07×10^{15}
0.395–0.400	1.24×10^{15}	1.30×10^{15}
0.400–0.405	1.46×10^{15}	1.54×10^{15}
0.405–0.410	1.64×10^{15}	1.72×10^{15}

18. How would you expect the ratio of Ca^{2+} to $CaCO_3$ in Reaction (5.9) (i.e., the dissolving of limestone in a weak solution of carbonic acid (H_2CO_3), to change with (a) decreasing temperature, and (b) increasing concentration of CO_2?

19. Figure A.1 summarizes, in a schematic diagram, the series of steps that oxidize CH_4 to CH_3OOH in the absence of NO_x. Use a similar diagram to show how, in the presence of NO_x, CH_3O_2 can be transformed back to OH. What input species are required, and what products are produced in this return portion of the cycle?

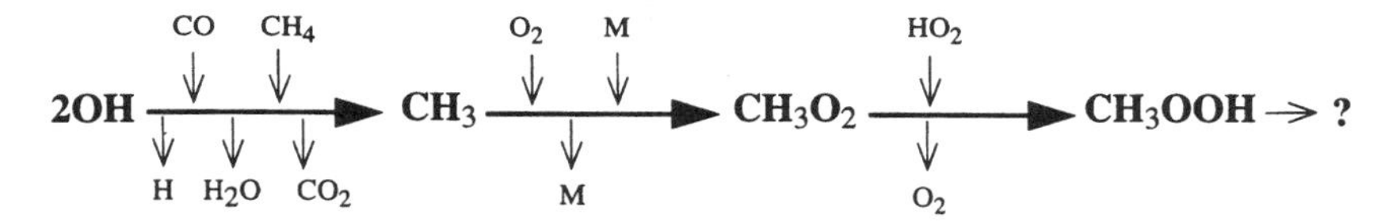

Figure A.1. Schematic diagram of steps that oxidize CH_4 to CH_3OOH in the absence of NO_x.

20. Assuming that SO_2 over the continents is confined to a layer of the atmosphere extending from the surface of the Earth up to a height of 5 km, and that the average deposition velocity of SO_2 onto the ground is $0.8\,cm\,s^{-1}$, how long would it take for all the SO_2 in this column to be deposited on the ground if all sources of SO_2 were suddenly switched off? Compare your answer with the residence time of SO_2 given in Table 2.1.

*21. In the atmosphere, NO reacts with O_3 to produce NO_2 and O_2. Nitric oxide also reacts with the hydroperoxyl (HO_2) radical to produce NO_2 and the hydroxyl radical (OH). In turn, NO_2 is photolyzed rapidly to produce NO and atomic oxygen. The atomic oxygen quickly combines with O_2 (when aided by an M) to produce O_3.

(a) Write balanced chemical equations to represent these four chemical reactions.

(b) Write differential equations to represent the time dependencies of the concentrations of each constituent (NO, O_3, NO_2, HO_2, OH, and O) in terms of appropriate constituent concentrations and rate coefficients.

(c) Neglecting hydroxyl-hydroperoxyl chemistry, and assuming steady-state conditions, derive an expression for the concentration of O_3 in terms of the concentrations of NO_2 and NO and appropriate rate coefficients.

*22. In the atmosphere the rate limiting step for the formation of carbon monoxide from methane is the oxidation of methane to CH_3 by the hydroxyl radical. Carbon monoxide, in turn, is oxidized to CO_2, which is also rate limited by a simple bimolecular reaction with OH. Methane is well mixed in the atmosphere, and its molecular fraction is ~2 ppm; its mean residence time is ~4 a. Carbon monoxide is reasonably well mixed in the northern hemisphere with a molecular fraction of ~0.1 ppm. The rate coefficient for the first oxidation is ~0.02 $ppm^{-1} s^{-1}$.

(a) Write down balanced chemical reactions for the rate limiting step of each oxidation.

(b) What is the mean residence time in the atmosphere of the CO molecule?

(c) What is the mean atmospheric concentration of the OH radical?

23. In the marine boundary layer, the concentration of nonseasalt sulfate (NSS) is ~0.50 $\mu g\, m^{-3}$. The concentration of NSS in rainwater is ~5.0 × 10^{-6} moles per liter. If the average annual rainfall is 2.0 m, the boundary layer depth ~3.0 km, and aerosol sulfate is the main source of NSS in the rainwater, estimate the average residence time of NSS sulfate in the marine boundary layer air. (Atomic weights of S and O are 32 and 16, respectively.)

*24. The world's oceans emit dimethyl sulfide (DMS) to the marine atmosphere. This (water) insoluble gas is then oxidized into acidic sulfate aerosols (which can be represented by SO_x) that are, in turn, deposited onto the oceans by rain. These two processes roughly establish steady-state mixing fractions for DMS and SO_x in marine air. For this problem assume DMS and SO_x are well mixed in the vertical and that there are three molecules of DMS per 10^{11} molecules of N_2 and O_2, and that the residence times of DMS and SO_x in the atmosphere are 1 and 3 days, respectively. The onshore flow of marine air brings DMS and SO_x inland, where DMS is similarly oxidized and SO_x is deposited onto the ground. Assume that the onshore winds are constant with height at 100 km per day, and that no neutralizing cations (e.g., NH_4^+) accompanies the SO_x aerosol, and that 1 m of rain falls per year over both oceans and continents.

(a) What is the steady-state mixing fraction of SO_x over the oceans?

(b) What is the rainborne flux of sulfur to the surface (in kg $m^{-2} yr^{-1}$) at the upwind shoreward edge of a narrow continent? (Assume that the scale height of the atmosphere with respect to air density is 8 km, the density of air at the surface 1.2 kg m^{-3}, the molecular weight of sulfur 32, and the apparent molecular weight of air 29.)

(c) Neglecting other sources of acidity, what is the average pH of the rain reaching the ground in (b)? (Assume that in rainwater $SO_3(aq) + H_2O((l) \rightarrow 2H^+(aq) + SO_4^=(aq)$.)

(d) What is the total flux of sulfur (in kg(S) a^{-1} per meter of coastline) to the surface of a wide continent?

(e) Show on a sketch how the mixing fractions of DMS and SO_x change in moving from 200 km offshore over the ocean to 500 km from the coastline over the land. Assume the wind is blowing onshore. Estimate (to ~10%) the downwind distances from the coastline at which DMS and SO_x fall to exp(−1) of its concentration in the middle of the ocean.

25. The following three reactions represent a gross simplification of the equilibrium chemistry of NO, NO_2, and O_3 in the troposphere

$$NO_2 + h\nu \xrightarrow{j} NO + O_2 \quad \text{(i)}$$

$$O + O_2 + M \xrightarrow{k_1} O_3 + M \quad \text{(ii)}$$

$$O_3 + NO \xrightarrow{k_2} NO_2 + O_2 \quad \text{(iii)}$$

Each of these reactions is rapid, and NO_2 is formed by Reaction (iii) as rapidly as it is depleted by Reaction (i). Derive an expression for the ratio of the equilibrium concentration of NO_2 to NO in terms of k_2, j, and the concentration of O_3.

26. The concentration of ozone just above the Earth's surface is 0.04 ppmv, and the rate coefficients for Reactions (i) and (iii) in Exercise 21 are $j = 4 \times 10^{-3}\,s^{-1}$ and $k_2 = 1 \times 10^{-20}\,m^3\,s^{-1}$ molecule^{-1}, respectively. Use the result from Exercise 25 to determine the ratio of the concentration of NO_2 to NO at 20°C and 1 atm.

27. If the aerosol number distribution is given by Eq. (6.3), derive an expression for dN/dD.

28. If the aerosol number distribution is given by Eq. (6.3), show that small fluctuations in the value of β about values of 2 and 3 will produce stationary values in the surface and volume distribution plots, respectively.

*29. Some of the dust particles that enter the air break down into smaller particles. If f is the fraction of dust particles that break down, and if m is the number of fragments into which a dust particle breaks, derive an expression for the volume distribution of dust particles formed by the breakdown, $n(v)$, in terms of the original volume distribution of dust that entered the atmosphere, $n_0(v')$. Assume f and m are independent of particle size.

30. Fresh water contains ~0.006% by mass of dissolved materials. Assuming that a bubble that bursts is fresh water and ejects the same mass of water into the air as it does in sea water, and that in both cases all of this water evaporates, compare the mass of material injected into the atmosphere by the bursting of a bubble in fresh water with that produced by the bursting of a bubble in sea water. (Amount of dissolved material in ocean water is ~0.2%.)

31. Assuming that Eq. (6.10) can be written as

$$-\frac{dN}{dt} = KN^2$$

where K is a constant for monodispersed particles of a given diameter and for a given temperature and atmospheric pressure ($K = 14 \times 10^{-16}\,\mathrm{m^3\,s^{-1}}$ for $0.10\,\mu\mathrm{m}$ diameter particles at 20°C and 1 atm), estimate the time required at 20°C and 1 atm for collisions to decrease the concentration of a monodispersed atmospheric aerosol with a particle diameter of $0.10\,\mu\mathrm{m}$ to one-half of its initial concentration of $1.0 \times 10^{11}\,\mathrm{m^{-3}}$. Assume that every collision results in coagulation.

*32. If the rate of decrease of particle number concentrations with height in the lower troposphere is $5 \times 10^{-3}\,\mathrm{cm^{-3}\,m^{-1}}$ for particles with diameters ~$1\,\mu\mathrm{m}$, calculate the rate of loss of these particles per unit volume of air due to sedimentation. (Take the density of the particles to be $2\,\mathrm{g\,cm^{-3}}$ and the dynamic viscosity of air to be $2 \times 10^{-5}\,\mathrm{N\,s\,m^{-2}}$.)

33. The amount of solar radiation reflected by a cloud depends on its optical depth (τ), which is given by

$$\tau = 2\pi h N (\bar{r})^2$$

where h is the cloud depth, N the total number concentration of droplets, and $\bar{r}$ the mean droplet radius.

(a) Show that

$$\tau = 2.4\left(\frac{W}{\rho_L}\right)^{2/3} h\, N^{1/3}$$

where W is the mass of cloud water per unit volume of air and ρ_L is the density of liquid water.

(b) If W and h are constant, show that

$$\frac{\Delta\tau}{\tau} = \frac{1}{3}\frac{\Delta N}{N}$$

(c) If the fraction of the total incident solar radiation reflected back into space in all directions by a cloud (called the *reflectivity* or *albedo*, A, of the cloud) is given by

$$A = \frac{\tau}{\tau + 6.7}$$

and if W and h are constant, show that

$$\frac{\Delta A}{\Delta N} = \frac{A(1-A)}{3N}$$

(d) If N is constant, for what value of A does $\Delta A/\Delta N$ have a maximum value?

(e) Sketch the general shape of a three-dimensional surface showing how $\Delta A/\Delta N$ varies with both N and A for values of A from 0 to 0.8 and values of N from 10 to $1{,}000\,\text{cm}^{-3}$. (For example, let $x = A$, $y = N$ and $z = \Delta A/\Delta N$, and sketch z as a function of x and y.)

(f) What is the approximate numerical maximum value of $\Delta A/\Delta N$ for $N = 10\,\text{cm}^{-3}$.

(g) About how many kilograms of CCN material would be required to produce a change in CCN concentrations of $1\,\text{cm}^{-3}$ from the surface of the earth up to a height of 1 km over the entire globe? Assume that each CCN has a mass of 10^{-19} kg, and that the Earth is a sphere of radius 6.37×10^{6} m.

(h) Assuming an atmospheric residence time of 2 days for CCN, what increase in the production rate (or flux) of CCN in the atmosphere would be required to produce the increase in the mass of CCN you have calculated in (g)?

(i) Assuming no other SO_2 loss process, what minimum percentage increase in the rate of anthropogenic SO_2 emissions would be required to achieve the increase in the rate of CCN production you have calculated in (h)? Assume that the increase in CCN mass is in the form of $(NH_4)_2SO_4$, and that the worldwide anthropogenic emission rate of SO_2 is about $150\,\mathrm{Tg\,a^{-1}}$.

34. What is the peak value of the supersaturation (S_c) of the Köhler curve for a NaCl particle of mass 10^{-19} kg?

35. The solubility of CO_2 in the surface layers of a lake with a temperature of 25°C and a pH of 5 is 1.25×10^{-5} M. Assuming that CO_2 in air and in the lake behaves as an ideal solution, calculate the partial pressure of CO_2 in a layer of air just above the surface of the lake. (*Hint*: See Figure 7.5.)

*36. Use Eq. (7.18) to calculate the concentration of CO_2 in an aqueous solution at 25°C with a pH of 8 that is in equilibrium with 330 ppm of CO_2 in air. Hence, show that, even at this high pH, there are about 2.5×10^4 molecules of CO_2 in the air for every CO_2 molecule in the cloud water. Assume a cloud liquid water content of 1 gram per cubic meter of air. (The successive acid dissociation constants for carbonic acid are 4.4×10^{-7} and 4.7×10^{-11}, respectively. The Henry's law constant for CO_2 at 25°C is $3.4 \times 10^{-2}\,\mathrm{M\,atm^{-1}}$. The total number of molecules in $1\,\mathrm{m^3}$ of air at 25°C is 2.5×10^{25}.)

*37. (a) Determine the value of the Henry's law constant for a gas that is equally distributed (in terms of mass) between air and cloud water if the liquid water content of the cloud is $1\,\mathrm{g\,m^{-3}}$ and the temperature 5°C. You may assume that the gas does not dissociate or react in water. (1 mole of a gas at atmospheric pressure and 5°C occupies a volume of 22.8 liters.)

(b) Using Figure 7.4, identify some gases that have Henry's law constants exceeding the critical value you have calculated in (a).

38. (a) From Eqs. (7.20)–(7.22) derive expressions for $[SO_2{\cdot}H_2O(aq)]$, $[HSO_3^-(aq)]$ and $[SO_3^{2-}(aq)]$ in terms of $k_H(SO_2)$, p_{SO_2}, K_{a1}, K_{a2}, and $[H_3O^+(aq)]$.

(b) Plot the concentrations of these three sulfur species, and their sum (i.e., $[S(IV)_{tot}]$), as a function of the pH of the solution for a partial pressure of $SO_2(g)$ of 1 ppbv.

(c) Plot as a function of pH the [S(IV)] mole fractions of the three sulfur species $\left(\text{i.e., } \frac{[SO_2 \cdot H_2O(aq)]}{[S(IV)]_{tot}}, \text{etc.}\right)$.

(d) What are the dominant forms of S(IV) in the solution for pH < 2, 3 < pH < 6, and pH > 7?

*39. Using Eq. (7.25) show that the rate of conversion of S(IV) to S(VI) by H_2O_2 in an aqueous solution is essentially independent of the pH of the solution.

40. With reference to Exercise 7.4, and the solution to that exercise, answer the following questions.

(a) What is the percentage rate of decrease of $SO_2(g)$ in terms of R, L, T, R_c^*, and the volume mixing ratio of $SO_2(g)$ in ppb (ξ_{SO_2})?

(b) If the S(IV) and A(aq) both obey Henry's law, derive an expression for the percentage rate of decrease of $SO_2(g)$ in terms of the effective Henry's law coefficient for $SO_2(g)$ (i.e., $k_{eff}(SO_2)$), the Henry's law coefficient for $A(g)$ (i.e., $k_H(A)$), L, R_c^*, T, k, and the volume mixing ratio of $A(g)$ in ppbv (i.e., ξ_A).

(c) If the aqueous-phase reaction rate given by R is $0.50\,\mu M\,s^{-1}$, $L = 0.30\,g\,m^{-3}$, $T = 278°K$, $R_c^* = 0.0821\,L\,atm\,deg^{-1}\,mol^{-1}$, and $\xi_{SO_2} = 5.0$ ppbv, what is the percentage rate of decrease of $SO_2(g)$ in the fog?

*41. It is conceivable that the increase in atmospheric CO_2 over, say, the past 50 years is due to an increase in the average temperature of the oceans, which would cause a decrease in the solubility of CO_2 in the oceans and therefore release CO_2 into the atmosphere. Estimate the percentage change in the CO_2 content of the atmosphere due to an average warming of 0.5°C in the upper (mixed) layer of the world's oceans over the past 50 years. (Assume that the average temperatures of the mixed layer of all the oceans has increased from 15.0°C to 15.5°C. You may treat the ocean water as pure water.)

Based on your calculation, does it appear likely that the measured increase in atmospheric CO_2 over the past 50 years is due to warming of the oceans?

You will need to use the following information.

The solubility, C_g, of a gas in a liquid is given by Henry's law:

$$C_g = k_H \, p_g$$

where k_H is the Henry's law constant, and p_g the partial pressure of the gas over the solution. For CO_2 in pure water, $k_H = 4.5 \times 10^{-2}\,\mathrm{M\,atm^{-1}}$ at 15°C.

The temperature dependence of k_H is given by

$$\ln\frac{k_H(T_2)}{k_H(T_1)} = \frac{\Delta H}{R^*}\left(\frac{1}{T_1} - \frac{1}{T_2}\right)$$

where for CO_2 in water $\Delta H = -20.4 \times 10^3\,\mathrm{J\,mol^{-1}}$, and R^* is the universal gas constant ($8.31\,\mathrm{J\,deg^{-1}\,mol^{-1}}$).

The total mass of carbon in the form of CO_2 in the mixed layer of the world's oceans is ~6.7×10^5 Tg, which is about the same as the mass of CO_2 in the atmosphere.

42. (a) The flux of N_2O into the atmosphere from natural sources is ~8 Tg(N) a^{-1}, the rate of increase of N_2O in the atmosphere is ~3 to 5 Tg(N) a^{-1}, and the rate of loss of N_2O (by chemical breakdown in the stratosphere) is ~9 Tg(N) a^{-1}. Use this information to calculate the possible range of values (in Tg(N) a^{-1}) of anthropogenic fluxes of N_2O into the atmosphere.

(b) By how much must anthropogenic emissions of N_2O be reduced to achieve a balanced N_2O budget in the atmosphere (i.e., no accumulation or deficit).

(c) If anthropogenic emissions of N_2O were suddenly reduced by the amounts calculated in (b), how long do you estimate it would take for the atmospheric N_2O budget to reach a balance?

43. Using the estimates given in Figure 8.1, calculate the residence times of (a) CO and (b) CO_2 in the troposphere. (c) Compare your calculated residence times with those given in Table 2.1. How do you account for any differences?

44. Estimate the residence time of aerosol sulfate in the marine boundary layer off the Washington coast given the following information for this region. Non-sea-salt sulfate (NSS) concentration = $0.50\,\mu\mathrm{g\,m^{-3}}$, average annual rainfall = 2.0 m, boundary layer depth = 3 km, and concentration of NSS in rainwater (due to aerosol sulfate) = 5×10^{-6} moles (of

sulfate) per liter. Assume that the sulfate is removed only by precipitation.

45. On Io, a moon of Jupiter, volcanoes emit copious sulfur vapor. Like oxygen, sulfur exists in several isometric forms, including S, S_2, and S_3. Assume that in Io's atmosphere sulfur chemistry is confined to these three sulfur species and that their interactions are limited to: photodissociation of the dimer (S_2); termolecular recombination of the atom (S) with the dimer and with an inert species ($M = S + S_2 + S_3$); and, bimolecular Reaction of the atom with the trimer (S_3).
 (a) Write balanced chemical equations for these three processes.
 (b) Write differential equations for the rates of formation of each of the three isomers, defining appropriate rate coefficients.
 (c) Assuming steady-state conditions and that the molecular fraction of S_2 greatly exceeds S or S_3, derive an expression for [S]/[M].
 (d) Sketch [S]/[M] as a function of altitude (z) on Io.

*46. The balanced chemical equation for the complete combustion of a general hydrocarbon fuel C_xH_y is given by

$$C_xH_y + \left(x + \frac{y}{4}\right)O_2 \rightarrow xCO_2 + \frac{y}{2}H_2O$$

 (a) If 1.0 mole of C_xH_y is completely burned, show that $3.7[x + (y/4)]$ moles of (unreacting) nitrogen will be contained in the emissions. Hence, write an expression, in terms of x and y, for the total number of moles of gases in the emissions.
 (b) In reality, combustion in cars converts all of the hydrogen in the fuel to H_2O and all of the carbon in the fuel to varying amounts of CO_2 and CO depending on the availability of oxygen.

 If a fraction f of the C_xH_y fuel is provided in excess of that required for complete combustion, derive an expression in terms of f, x, and y for the mole fraction of CO in the emissions (i.e., the ratio of the number of moles of CO to the total number of moles in the emissions). Assume that oxygen is made available to the fuel at the rate required for complete combustion (even though complete combustion is not achieved), and that the only effect of the excess C_xH_y is to add CO to the emissions and to change the amount of CO_2 emitted.

(c) Assuming that CH_2 is a reasonable approximation for a general hydrocarbon fuel, use the result from (b) to determine the concentrations (in ppmv and percent) of CO in the emissions from an engine for the following values of f: 0.0010, 0.010, and 0.10.

47. Assuming only photochemical equilibrium and starting with Reactions (10.1)–(10.4), derive expressions for the rates of change of the concentrations of atomic oxygen (n_1) and O_2 (n_2) in terms of the concentrations of O_3 (n_3) and the inert molecule (n_M), and the rate constants for Reactions (10.1)–(10.4).

48. Using the expressions for dn_1/dt, dn_2/dt, and dn_3/dt derived in Exercise 10.1 (in Chapter 10) and Exercise 47, show that $n_1 + 2n_2 + 3n_3 =$ constant. Why could you have predicted this result?

49. If the following elementary reactions are responsible for converting ozone into molecular oxygen

$$O_3 \rightleftarrows O_2 + O \qquad \text{(i)}$$

$$O_3 + O \rightleftarrows 2O_2 \qquad \text{(ii)}$$

what is (a) the overall chemical reaction, (b) the intermediate, (c) the rate law for each elementary reaction, and (d) the rate controlling elementary reaction if the rate law for the overall reaction is

$$\text{Rate} = k[O_3]^2[O_2]^{-1}$$

where k is a rate coefficient, and (e) on what would you surmise [O] depends?

50. Reaction (10.3) consists of the following elementary steps

$$O + M \underset{k_2}{\overset{k_1}{\rightleftarrows}} OM$$

$$O_2 + OM \xrightarrow{k_3} O_3 + M$$

$$\text{NET:} \quad O + O_2 + M \xrightarrow{k_b} O_3 + M$$

(a) Write expressions for $d[\mathrm{OM}]/dt$ and $d[\mathrm{O_3}]/dt$.
(b) Assuming $d[\mathrm{OM}]/dt = 0$, derive an expression for [OM] in terms of k_1, k_2, k_3, [O], and [M]. (Assume a constant mixing fraction x for atmospheric O_2.)
(c) Solve for k_b in terms of k_1, k_2, k_3, [M], and x. Sketch log k_b versus log [M].

*51. In the middle and upper stratosphere, O_3 concentrations are maintained at roughly steady values by a number of chemical reactions. For the purpose of this problem, you may assume that at around a temperature of 220 K

$$\frac{dX}{dt} = k_1 - k_2 X^2$$

where

$$\mathrm{X} = \frac{\text{concentration of } \mathrm{O_3} \text{ molecules}}{\text{concentration of all molecules}}$$

$$k_1 = (\text{constant})\exp\left(\frac{300}{T}\right)s^{-1}$$

$$k_2 = 10.0\exp\left(\frac{-1{,}100}{T}\right)s^{-1}$$

(a) Doubling the concentration of CO_2 in the atmosphere is predicted to cool the middle stratosphere by about 2°C. What fractional change in X would you expect from this temperature perturbation?
(b) If X were temporarily raised by 1.0% above its steady-state value of 5.0×10^{-7}, how long would it take for this perturbation to fall to exp(−1) of 1.0% at 220 K? (exp 1 = 2.7)

52. Write down the catalytic cycles and the net reactions corresponding to Reactions (10.12) when (a) X = H, and (b) X = OH.

53. A variation on the catalytic reaction cycle (10.24) is

$$\mathrm{ClO + ClO + M \xrightarrow{k_1} (ClO)_2 + M} \qquad \text{(i)}$$

$$\mathrm{(ClO)_2} + h\nu \xrightarrow{j_2} \mathrm{Cl + ClOO} \qquad \text{(iia)}$$

$$\mathrm{(ClO)_2 + M \xrightarrow{k_2} 2ClO + M} \qquad \text{(iib)}$$

$$\mathrm{ClOO + M \xrightarrow{k_3} Cl + O_2 + M} \qquad \text{(iii)}$$

$$\mathrm{2Cl + 2O_3 \xrightarrow{k_4} 2ClO + 2O_2} \qquad \text{(iv)}$$

In this cycle, two possible fates for $(ClO)_2$ are indicated: photolysis to form Cl and ClOO (Reaction (iia)), in which case Reactions (iii) and (iv) follow, or thermal decomposition to produce ClO (Reaction (iib)).

(a) What is the net effect of this cycle if Reaction (iia) dominates? What is the net effect if Reaction (iib) dominates?

(b) Derive equations for the rate of change of O_3, Cl, $(ClO)_2$, and ClOO, assuming that Reaction (iia) dominates and that Reaction (iib) can be neglected.

(c) Assume that the concentrations of Cl, $(ClO)_2$, and ClOO are in steady state, and using the results from part (b), derive an expression for the concentration of Cl in terms of the concentrations of ClO, O_3, and M, and the values of k_1 and k_4.

(d) Use the results of parts (b) and (c) to find an expression for the rate of change in the concentration of O_3 in terms of k_1 and the concentrations of ClO and M.

(e) If removal of O_3 by chlorine chemistry becomes a significant part of the O_3 budget and total chlorine increases linearly with time, what mathematical form do you expect for the time dependence of the O_3 concentration (e.g., linear in time, square-root of time, etc.)?

*54. In the atmosphere at altitudes near and above 30 km the following reactions significantly affect the chemistry of ozone

$$\mathrm{O_3 + h\nu \xrightarrow{j_1} O_2 + O^*} \qquad (j_1 = 1\times10^{-4}\,\mathrm{s^{-1}})$$

$$\mathrm{O^* + M \xrightarrow{k_1} O + M} \qquad (k_1 = 1\times10^{-11}\,\mathrm{cm^3\,s^{-1}})$$

$$\mathrm{O^* + H_2O \xrightarrow{k_2} HO + HO} \qquad (k_2 = 2\times10^{-6}\,\mathrm{cm^3\,s^{-1}})$$

$$\mathrm{O + O_2 + M \xrightarrow{k_3} O_3 + M} \qquad (k_3 = 6\times10^{-34}\,\mathrm{cm^6\,s^{-1}})$$

$$\mathrm{HO + O_3 \xrightarrow{k_4} HO_2 + O_2} \qquad (k_4 = 2\times10^{-14}\,\mathrm{cm^3\,s^{-1}})$$

$$\mathrm{HO_2 + O_3 \xrightarrow{k_5} HO + O_2 + O_2} \qquad (k_5 = 3\times10^{-16}\,\mathrm{cm^3\,s^{-1}})$$

$$\mathrm{HO + HO_2 \xrightarrow{k_6} H_2O + O_2} \qquad (k_6 = 3\times10^{-11}\,\mathrm{cm^3\,s^{-1}})$$

where O^* is an electronically excited metastable state of atomic oxygen. The free-radical species HO and HO_2 are collectively labeled "odd hydrogen." At 30 km the molecular density of the atmosphere is about $5 \times 10^{17}\,cm^{-3}$, and the molecular fractions of water vapor and O_3 are each about 2×10^{-6} and that of oxygen is 0.2.

(a) What are the approximate steady-state molecular fractions of O^*, HO, and HO_2?

(b) What is the approximate mean lifetime of odd hydrogen under steady-state conditions?

(c) For every odd-hydrogen entity produced by the k_2 process, about how many O_3 molecules are destroyed under steady-state conditions?

(*Hint*: The steps associated with k_4 and k_5 occur many times for each formation or loss of odd hydrogen.)

Appendix II

Answers to exercises in Appendix I and hints and solutions to the more difficult exercises

1. (q) $5 \times 10^6\,cm^{-3}$
   (t) From about 0.6 to 1.7 km.
   (u) About 0.7 km.
   (z) The oxidation states of sulfur in H_2S, DMS (CH_3SCH_3), COS, and CS_2 are all –2. The oxidation state of sulfur in $SO_4^=$ is +6.
2. Balanced series of reactions is

$$\left.\begin{array}{c} 7H_2O + h\nu \rightarrow 7OH + 7H \\ 3OH + 3OH \rightarrow 3O + 3H_2O \\ O + OH \rightarrow O_2 + H \\ O + O + M \rightarrow O_2 + M \end{array}\right\}$$

$$\text{Net:}\quad 7H_2O + h\nu \rightarrow 8H + 3H_2O + 2O_2$$

or

$$2H_2O + h\nu \rightarrow 4H + O_2$$

Two water molecules are needed to produce one O_2 molecule

3. The net balanced reaction is $2H_2O + h\nu \rightarrow 4H + O_2$.
   No.
4. (a) About 108 bar at 300 K and 5.6×10^{-6} bar at 400 K.
   (b) No.
5. (a) Yes. Because the amount of methane produced would be determined by the bacterial amounts and hydrogen produced in the local environments (not by hydrogen con-

centrations in the atmosphere as in the mechanism discussed in Exercise 4).

(b) 164.9 kJ is released per mole of CH_4 produced.

*6. By the year 2000 the mass of carbon entering the atmosphere/year will be

$$3 \times 10^{13} \times \frac{80}{100} \times \frac{1}{2} = 12 \times 10^{12}\,\text{kg}$$

Since the molecular weights of CO_2 and carbon-12 are 44 and 12, respectively, the mass of CO_2 formed is

$$\frac{44}{12}(12 \times 10^{12})\,\text{kg}$$

The mass of a CO_2 molecule = (44) m_H, where m_H = mass of hydrogen atom. Therefore,

Number of CO_2 molecules added to atmosphere/year

$$= \frac{\frac{44}{12}(12 \times 10^{12})}{44 m_H} = \frac{10^{12}}{m_H}$$

Mass of the atmosphere
= (number of molecules in the atmosphere)
× (apparent molecular weight of air) m_H

Therefore,

$$\text{Number of molecules in atmosphere} = \frac{5.1 \times 10^{18}}{(29) m_H}$$

Hence,

$$\frac{\text{Number of } CO_2 \text{ molecules added per year}}{\text{Number of molecules in atmosphere}}$$

$$= \frac{\dfrac{10^{12}}{m_H}}{\dfrac{5.1 \times 10^{18}}{(29) m_H}}$$

$$= \frac{29}{5.1} 10^{-6} = 5.7 \times 10^{-6}$$

Since the volume occupied by a gas is proportional to the number of molecules in the gas, the annual rate of increase in CO_2 in the year 2000 will be 5.7 ppmv

7. $\frac{1}{k[\mathrm{A}]_0}$

8. 1.3 g
9. (a) 1 h
   (b) 20 h
10. 0.599 ppbv
11. 45 ppbv of NO_2
12. *Hint*: Use Eq. (3.2).
13. 98%
14. 0.51
15. 8.4 km

    From Eq. (3.2) we see that the scale height (H) of the atmosphere, with respect to density, is the height above the Earth's surface where the air density is exp (−1), or about 37%, of the density at the surface.
16. A plot of ln E_λ (on the ordinate) versus sec θ (on the abscissa) has an intercept on the ordinate of ln $E_{\lambda\infty}$ and a slope of $-\tau_\lambda$.
17. (a) $6.60 \times 10^{-3}\,\mathrm{s}^{-1}$
    (b) $7.08 \times 10^{-3}\,\mathrm{s}^{-1}$
    (c) On 1 August the solar zenith angle is lower; therefore, the Sun is higher in the sky, and the path length through the atmosphere is less, than on 1 March. Hence, more photons reach the Earth's surface on 1 August.
    (d) $3.3 \times 10^{13}\,\mathrm{m}^{-3}\,\mathrm{s}^{-1}$
18. (a) Ca^{2+} would increase as temperature is lowered because more CO_2(g) dissolves in water.
    (b) Ca^{2+} would increase with increasing CO_2(g) because of more H_2CO_3(aq).
19. $$CH_3O_2 \xrightarrow[\downarrow\ NO_2]{NO\ \downarrow} CH_3O \xrightarrow[\downarrow\ HCHO]{O_2\ \downarrow} HO_2 \xrightarrow[\downarrow\ NO_2]{NO\ \downarrow} OH$$
20. Seven days, compared to 3.7 days given in Table 2.1. This calculation gives an upper limit to τ because there are other (*in situ*) removal mechanisms.

*21. (a) $$NO + O_3 \xrightarrow{k_1} NO_2 + O_2$$

$$NO + HO_2 \xrightarrow{k_2} NO_2 + OH$$

$$NO_2 + h\nu \xrightarrow{j} NO + O$$

$$O + O_2 + M \xrightarrow{k_3} O_3 + M$$

(b) $$\frac{d[NO]}{dt} = -k_1[NO][O_3] - k_2[NO][HO_2] + j[NO_2]$$

$$\frac{d[O_3]}{dt} = -k_1[NO][O_3] + k_3[O][O_2][M]$$

$$\frac{d[NO_2]}{dt} = -\frac{d[NO]}{dt}$$

$$\frac{d[HO_2]}{dt} = -k_2[NO][HO_2]$$

$$\frac{d[OH]}{dt} = -k_2[NO][HO_2]$$

$$\frac{d[O]}{dt} = j[NO_2] - k_3[O][O_2][M]$$

(c) Under steady-state conditions

$$\frac{d[NO]}{dt} = 0$$

and, if $[HO_2] = 0$, the first equation in (b) becomes

$$O_3 = \frac{j[NO_2]}{k_1[NO]}$$

*22. (a) $$CH_4 + OH \xrightarrow{k_1} CH_3 + H_2O \qquad \text{(i)}$$

(Followed by the series of reactions discussed in Section 5.2 that lead to CO.)

$$CO + OH \xrightarrow{k_2} CO_2 + H \qquad \text{(ii)}$$

(b) The residence time (τ) for a monomolecular reaction is $1/k$.

For a bimolecular reaction, such as (i), we can define a pseudo first-order rate coefficient k_1 as $k_1[OH]$. Therefore,

$$\tau_{CH_4} = \frac{1}{k_1[OH]}$$

Hence,

$$k_1 = \frac{1}{4[OH]} \quad \text{(iii)}$$

Since (i) is the rate-limiting step for production of CO,

$$\left[\frac{d[CO]}{dt}\right]_{prod} = k_1[CH_4][OH] = \frac{1}{4}[CH_4]$$

The rate of destruction of CO is, from (ii),

$$\left[\frac{d[CO]}{dt}\right]_{dest} = k_2[CO][OH]$$

Hence,

$$\left[\frac{d[CO]}{dt}\right] = \frac{[CH_4]}{4} - k_2[CO][OH]$$

Under steady-state conditions

$$\frac{[CH_4]}{4} - k_2[CO][OH] = 0$$

or

$$\frac{1}{k_2[OH]} = \frac{4[CO]}{[CH_4]} = \frac{4(0.1)}{2} = 0.2\,\text{a}$$

But, from (ii), the residence time of CO molecules is

$$\tau_{CO} = \frac{1}{k_2[OH]} = 0.2\,\text{a}$$

(c) From (iii)

$$[OH] = \frac{1}{4k_1}$$

In this relation, the 4 has units of years, so if $k_1 = 0.02$ ppm^{-1} s^{-1}, we must change 4 years to seconds. Then,

$$[OH] = \frac{1}{(4 \times 365 \times 24 \times 60 \times 60)(0.02)}\,\text{ppm}$$
$$= 3.96 \times 10^{-7}\,\text{ppm} \simeq 0.4\,\text{ppt}$$

23. 14 h

*24. (a) For steady-state concentration of SO_x over the ocean

$$\frac{d[SO_x]}{dt} = (\text{Loss rate of DMS to } SO_x) - (\text{Loss rate of } SO_x \text{ to ground}) = 0$$

Therefore,

$$0 = \frac{[\text{DMS}]_{\text{total column}}}{\tau_{\text{DMS}}} - \frac{[SO_x]_{\text{total column}}}{\tau_{SO_x}}$$

Hence,

$$[SO_x]_{\text{total column}} = \frac{\tau_{SO_x}}{\tau_{\text{DMS}}}[\text{DMS}]_{\text{total column}}$$

Since both SO_x and DMS are well mixed vertically,

$$[SO_x]_{\text{total column}} \propto \text{mixing fraction of } SO_x = [SO_x]$$

$$[\text{DMS}]_{\text{total column}} \propto \text{mixing fraction of DMS} = [\text{DMS}]$$

Therefore,

$$[SO_x] = \frac{3}{1}(3 \times 10^{-11}) = 9 \times 10^{-11}$$

(b) Flux of sulfur (S) to surface $= \dfrac{[S]_{\text{total column}}}{\tau_{SO_x}}$

Also,

$$[S]_{\text{total column}} = \int_0^\infty \rho_s(z)dz$$

where, $\rho_s(z)$ = density of sulfur at z. Therefore,

$$[S]_{\text{total column}} = \int_0^\infty \left(\rho_{\substack{\text{air} \\ \text{at } z}}\right)[SO_x]\frac{M_s}{M_a}dz$$

where, M_s = molecular weight of sulfur and M_a = apparent molecular weight of air.

Or, since

$$\rho_{\text{air at } z} = \rho_{\substack{\text{air at} \\ \text{surface}}} \exp\left(-\frac{z}{H}\right)$$

where, H = scale height of air with respect to *density*,

$$[\mathrm{S}]_{\text{total column}} = [\mathrm{SO}_x]\frac{M_s}{M_a}\rho_{\substack{\text{air at}\\\text{surface}}}\int_0^{\infty}\exp\left(-\frac{z}{H}\right)$$

Therefore,

$$[\mathrm{S}]_{\text{total column}} = [\mathrm{SO}_x]\frac{M_s}{M_a}\rho_{\substack{\text{air at}\\\text{surface}}}H$$

Hence,

$$\text{Flux of sulfur to surface} = \frac{[\mathrm{SO}_x]\frac{M_s}{M_a}\left(\rho_{\substack{\text{air at}\\\text{surface}}}\right)H}{\tau_{\mathrm{SO}_x}}$$

$$= \frac{(9\times10^{-11})\frac{32}{29}(1.2)(8\times10^{3})}{\frac{3}{365}}$$

$$= 1.2\times10^{-4}\,\mathrm{kg\,m^{-2}\,a^{-1}}$$

(c) $\mathrm{pH} \equiv -\log\,[\mathrm{H}^+]$

where $[\mathrm{H}^+]$ is in moles/liter. Since,

$$\mathrm{SO_3(aq) + H_2O(l) \rightarrow 2H^+(aq) + SO_4^{=}(aq)}$$

two protons are produced for every sulfur atom in rain. Therefore,

$$[\mathrm{H}^+]_{\text{rain}} = 2[\mathrm{S}]_{\text{rain}}\frac{M_H}{M_S} = 2[\mathrm{S}]_{\text{rain}}\frac{1}{32}$$

But

$$[\mathrm{S}]_{\text{rain}}(\text{rain rate}) = \text{flux of sulfur to the surface}$$

$$\left\{\text{Units: }\frac{\text{mole}}{\text{L}}\left(\frac{10^{-3}\text{kg}}{\text{mole}}\right)\times\frac{\text{m}}{\text{a}}\left(\frac{1\text{m}^2}{\text{m}^2}\frac{10^3\text{L}}{\text{m}^3}\right) = \frac{\text{kg}}{\text{m}^2\text{a}}\right\}$$

Therefore,

$$[\mathrm{S}]_{\text{rain}}\times 1 = 1.2\times10^{-4}$$

Therefore,

$$[\mathrm{H}^+]_{\text{rain}} = 2(1.2\times10^{-4})\frac{1}{32} = 7.5\times10^{-6}\ \mathrm{M}$$

Hence,

$$\text{pH of rain reaching the ground} = -\log(7.5\times 10^{-6}) = 5.1$$

(d) For a wide continent, all of the DMS and SO_x from the ocean will be deposited on surface. Also, from Exercise 12,

Mass of air in atmosphere = (density of air at surface) × (scale height). Therefore,

$$\text{Mass of air in atmosphere} = \rho_{\substack{\text{air at}\\ \text{surface}}}(8\times 10^3)$$

$$\text{Mass of sulfur in air column (in units of kg (S) m}^{-2})$$

$$= \{[\text{DMS}]+[\text{SO}_\text{x}]\}\rho_{\substack{\text{air at}\\ \text{surface}}}(8\times 10^3)\frac{M_s}{M_a}$$

In one day the amount of sulfur flowing inland (in units of kg (S) m^{-1} day^{-1})

$$= \{[\text{DMS}]+[\text{SO}_\text{x}]\}\rho_{\substack{\text{air at}\\ \text{surface}}}(8\times 10^3)\times(100)(10^3)\frac{M_s}{M_a}$$

$$= (3\times 10^{-11}+9\times 10^{-11})(1.2)(8\times 10^3)\times 10^5\times\frac{32}{29}\times 365$$

$$= 46\ \text{kg(S)}$$

(e) See Figure A.2.

For DMS downwind distance from coastline to fall to exp(−1) = 100 km

For SO_x downwind distance from coastline to fall to exp(−1) = 400 km

(*Note*: Because DMS feeds SO_x for about 1 day inland, SO_x falls to exp(−1) after ~4 days, not 3 days.)

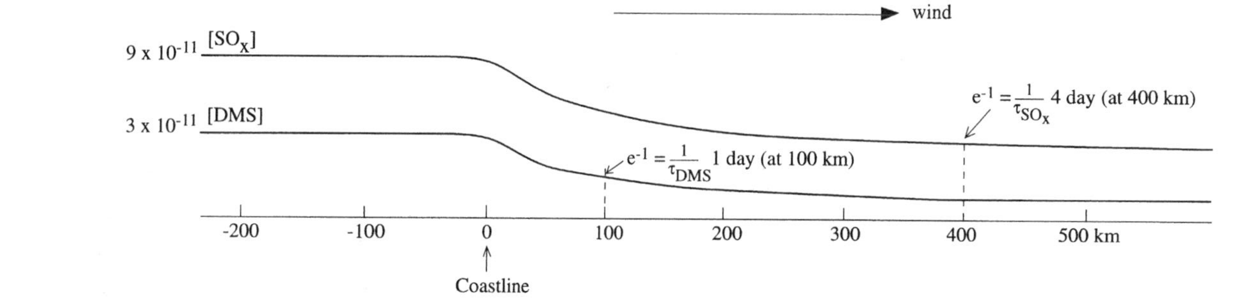

Figure A.2. Schematic diagram for solution to Exercise 24(e) showing mixing fractions of SO_x and DMS from 200 km offshore over ocean to 500 km from coastline over the land.

25.
$$\frac{[NO_2]}{[NO]} = \frac{k_2}{j}[O_3]$$

26. About 3.

27.
$$\frac{dN}{dD} = \frac{C_2}{\ln 10} D^{-(\beta+1)}$$

*29. If a fraction f of the original dust particles each break down into m particles, the volume v of the particles in the volume interval v to $v + dv$ formed by the breakdown of dust particles of volume v is

$$v = \frac{fv'}{m}$$

Therefore,

$$dv = \frac{f}{m} dv'$$

The concentrations of particles formed by the breakdown that have a volume between v and $v + dv$ is

$$n(v)dv = m\, n_0(v')dv'$$

Therefore,

$$n(v)\frac{f}{m} dv' = m\, n_0(v')dv'$$

or

$$n(v) = \frac{m^2}{f} n_0(v')$$

30. A fresh water bubble injects ~0.2% of the material ejected by a sea water bubble.

31. About 2h.

*32. Consider a slab of air at height z above the Earth's surface (Fig. A.3). Let the slab have a unit cross-sectional area and a thickness dz. Let particles of diameter D be present in number concentrations n at height z and $n + dn$ at height $z + dz$. Let v_s be the terminal settling velocity of these particles. Then, as a result of sedimentation, the number of particles of diameter D that pass through the upper surface of the slab in

1s is $(n + dn)\ v_s$, and the number that sediment out of the lower surface of the slab in 1s is nv_s. Therefore, the net number of particles of diameter leaving the slab in 1s is $nv_s - (n + dn)\ v_s = -v_s\ dn$. Since the volume of the slab is dz, the rate of loss of particles of diameter D per unit volume of air is $-v_s\ dn/dz$. Combining this with Eq. (6.13b) for the case when $\rho_p >> \rho$, we get for particles of diameter D.

Rate of loss per unit volume due to sedimentation

$$= -\frac{D^2 g \rho_p}{18\mu}\frac{dn}{dz}$$

Substituting $D = 1 \times 10^{-6}\,\text{m}$, $g = 9.82\,\text{m}\,\text{s}^{-2}$, $\rho_p = 2 \times 10^3\,\text{kg}\,\text{m}^{-3}$, $\mu = 2 \times 10^{-5}\,\text{N}\,\text{s}\,\text{m}^{-2}$ and $dn/dz = -5 \times 10^{-3}\,\text{cm}^{-3}\,\text{m}^{-1} = -5 \times 10^3\,\text{m}^{-3}\,\text{m}^{-1}$, we get for particles of diameter 1 μm.

Rate of loss per unit volume due to sedimentation

$$= \frac{(10^{-6})^2 (9.82)(2 \times 10^3)(5 \times 10^3)}{18(2 \times 10^{-5})}$$

$$\simeq 0.3\,\text{m}^{-3}\,\text{s}^{-1}$$

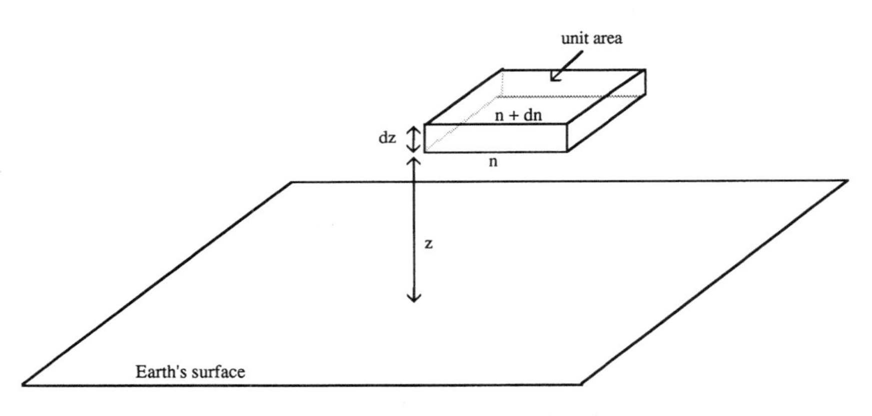

Figure A.3. Sketch for solution to Exercise 32.

33. (d) 0.5
(e) See Fig. A.4.

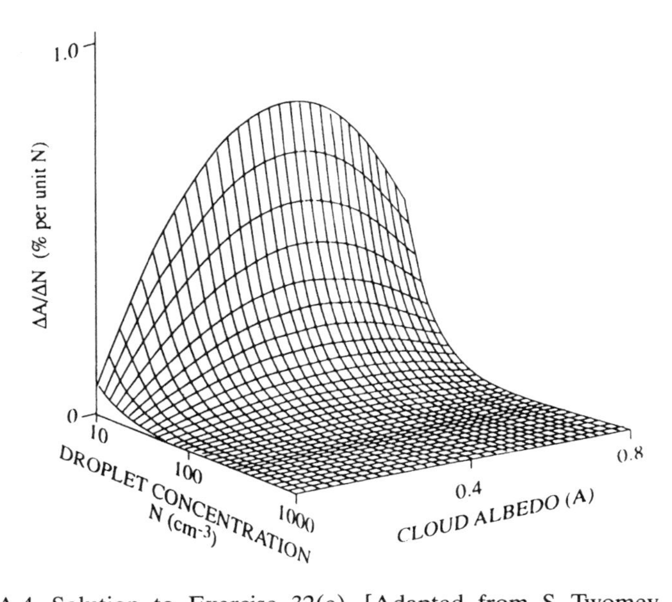

Figure A.4. Solution to Exercise 32(e). [Adapted from S. Twomey, *Atmos. Environ.*, **25A**, 2435. Copyright © 1991 with permission from Elsevier Science.]

(f) $0.008\,\text{cm}^{-3}$
(g) About 5×10^4 kg
(h) $9 \times 10^6\,\text{kg a}^{-1}$
(i) 0.003%

34. $S_c = 0.35\%$

35. 368 ppmv

*36. From Eq. (7.18)

$$[CO_2(aq)]_{tot} = k_H[CO_2]p_{CO_2}\left[1 + \frac{K_{a1}}{[H_3O^+(aq)]} + \frac{K_{a1}K_{a2}}{[H_3O^+(aq)]^2}\right]$$

$$= (3.4 \times 10^{-2})p_{CO_2}\left[1 + \frac{4.4 \times 10^{-7}}{[H_3O^+(aq)]} + \frac{4.4 \times 10^{-7} \times 4.7 \times 10^{-11}}{[H_3O^+(aq)]^2}\right]$$

where, p_{CO_2} is in atm.

$$p_{CO_2} = 330\,\text{ppm} = 330 \times 10^{-6}\,\text{atm}$$

$$\mathrm{pH} = -\log[\mathrm{H_3O^+(aq)}]$$

Therefore,

$$8 = -\log[\mathrm{H_3O^+(aq)}]$$

or

$$[\mathrm{H_3O^+(aq)}] = 10^{-8}\,\mathrm{M}$$

Therefore,

$$\begin{aligned}
[\mathrm{CO_2(aq)}]_{\mathrm{tot}} &= (3.4\times10^{-2})(330\times10^{-6})\\
&\quad\left[1+\frac{4.4\times10^{-7}}{10^{-8}}+\frac{4.4\times10^{-7}\times4.7\times10^{-11}}{(10^{-8})^2}\right]\\
&= 1,122\times10^{-8}\left[1+44+\times\frac{20.7\times10^{-18}}{10^{-16}}\right]\\
&= 1,122\times10^{-8}[1+44+0.21]\\
[\mathrm{CO_2(aq)}]_{\mathrm{tot}} &= 5\times10^{-4}\,\mathrm{M} = 5\times10^{-4}\ \text{moles/liter of water}
\end{aligned}$$

How does this compare to amount of CO_2 in the air?

$$\begin{aligned}
[\mathrm{CO_2(aq)}]_{\mathrm{tot}} &= 5\times10^{-14}\,\frac{\text{moles}}{\text{liter of cloud water}}\\
&= 5\times10^{-4}\times N_A\,\frac{\text{molecules}}{\text{liter of cloud water}}\\
&= 5\times10^{-4}\times6\times10^{23}\,\frac{\text{molecules}}{\text{liter of cloud water}}\\
&= 3\times10^{20}\,\frac{\text{molecules}}{\text{liter of cloud water}}
\end{aligned}$$

How many liters of air are required for 1 liter of cloud water? There is

$$1\,\mathrm{g}\ \text{of cloud water per}\ 1\,\mathrm{m}^3\ \text{of air}$$

or

$$10^{-3}\,\mathrm{g}\ \text{of cloud water per 1 liter of air}$$

Since 1 cm^3 of water has a mass of 1 g, 1 liter of water has a mass of 10^3 g.

Therefore,

1 gram of cloud water occupies a volume of 10^{-3} liter

or

10^{-3} gram of cloud water occupies a volume of 10^{-6} liter of water

Therefore, there are

10^{-6} liter of cloud water per 1 liter of air

or

1 liter of cloud water per 10^6 liter of air

Hence,

$$[CO_2(aq)]_{tot} = 3 \times 10^{20} \text{ molecules per } 10^6 \text{ liter of air}$$
$$= 3 \times 10^{14} \frac{\text{molecules}}{\text{liter of air}}$$

But there are 2.5×10^{25} total molecules in 1 m^3 of air, or 2.5×10^{22} total molecules in 1 liter of air.

Therefore,

Number of CO_2 molecules in 1 liter of air
$$= (330 \times 10^{-6}) \times 2.5 \times 10^{22}$$
$$= 8.25 \times 10^{18}$$

Therefore,

$$\begin{array}{l}\text{Ratio of number of molecules of} \\ CO_2 \text{ in water drops to that in air}\end{array} = \frac{3 \times 10^{14}}{8.25 \times 10^{18}}$$
$$= 3.6 \times 10^{-5}$$

Therefore,

$$\begin{array}{l}\text{Number of } CO_2 \text{ molecules in air for} \\ \text{every } 1\, CO_2 \text{ molecule in water}\end{array} = \frac{1}{3.6 \times 10^{-5}}$$
$$= 2.7 \times 10^4$$

*37. (a) Let the gas be X, then

$$\begin{array}{l}\text{Amount of X (in moles) in the cloud} \\ \text{water (per } m^3 \text{ of air)}\end{array} = [X(aq)] \frac{LWC}{\rho_w}$$

where LWC is the liquid water content (in $kg\,m^{-3}$), and ρ_w is the density of liquid water (in kg per liter).

Since

$$[X(aq)] = k_H(X)p(X)$$

$$\begin{matrix}\text{Amount of X (in moles) in} \\ \text{cloud water (per m}^3\text{ of air)}\end{matrix} = k_H(X)p(X)\frac{LWC}{\rho_w} \quad \text{(i)}$$

Since 1 mole of a gas at 1 atm and 5°C occupies a volume of 22.8 liters (or $0.0228\,m^3$), 1/0.0228 moles occupies a volume of $1\,m^3$ at 1 atm and 5°C. Therefore, if the partial pressure of X is $p(X)$ atm,

$$1\,m^3 \text{ of air contains } \frac{p(X)}{0.0228} \text{ moles of X at 5°C} \quad \text{(ii)}$$

For same amount of X in air and water, we have from (i) and (ii):

$$k_H(X)p(X)\frac{LWC}{\rho_w} = \frac{p(X)}{0.0228}$$

Therefore,

$$k_H(X) = \frac{\rho_w}{0.0228(LWC)}$$

where

$$\rho_w = 1\,g\,cm^{-3} = 1\text{ kg per liter}$$

$$LWC = 1\,g\,m^{-3} = 10^{-3}\,kg\,m^{-3}$$

Therefore,

$$k_H(X) = \frac{1}{0.0228(10^{-3})} = 4.4 \times 10^4\,M\,atm^{-1}$$

Hence, the required critical value of the Henry's law constant is $k_H(X) = 4.4 \times 10^4\,M\,atm^{-1}$.

(b) H_2O_2, NO_3

38. (a) $$[SO_2 \cdot H_2O(aq)] = k_H(SO_2)p_{SO_2}$$

$$\left[\mathrm{HSO_3^-(aq)}\right] = \frac{k_H(\mathrm{SO_2})K_{\mathrm{a1}}p_{\mathrm{SO_2}}}{\left[\mathrm{H_3O^+(aq)}\right]}$$

$$\left[\mathrm{SO_3^{2-}(aq)}\right] = \frac{k_H(\mathrm{SO_2})K_{\mathrm{a1}}K_{\mathrm{a2}}p_{\mathrm{SO_2}}}{\left[\mathrm{H_3O^+(aq)}\right]}$$

(b)

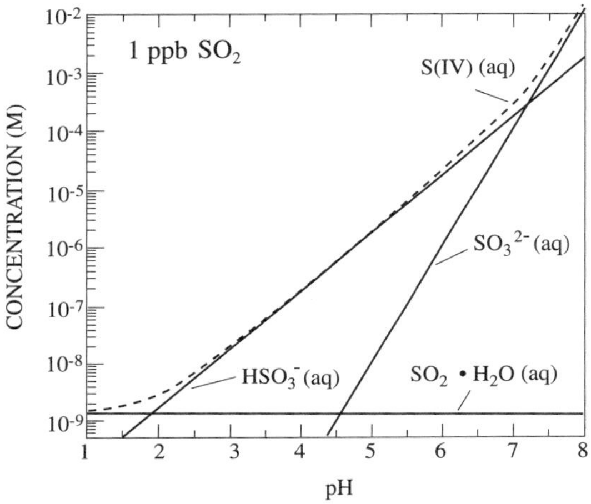

Figure A.5. Answer to Exercise 38(b). [Adapted from *Atmospheric Chemistry and Physics* by J. H. Seinfeld and S. N. Pandis. Copyright © 1998 John Wiley & Sons, Inc. Reprinted by permission of John Wiley & Sons, Inc.]

(c)

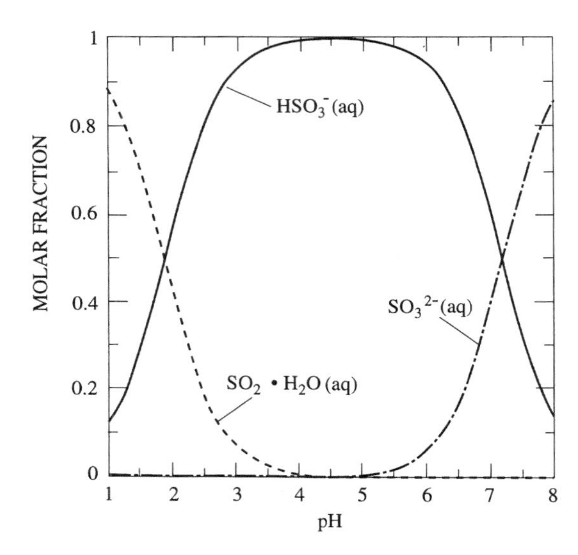

Figure A.6. Answer to Exercise 38(c). [Adapted from *Atmospheric Chemistry and Physics* by J. H. Seinfeld and S. N. Pandis. Copyright © 1998 John Wiley & Sons, Inc. Reprinted by permission of John Wiley & Sons, Inc.]

(d) For pH < 2, $SO_2 \cdot H_2O$(aq) is dominant.
For 3 < pH < 6, HSO_3^-(aq) is dominant.
For pH > 7, SO_3^{2-}(aq) is dominant.

*39. From Eq. (7.25)

$$-\frac{d[\mathrm{S(IV)}]}{dt} = \frac{k[\mathrm{H_3O^+(aq)}][\mathrm{H_2O_2(aq)}][\mathrm{HSO_3^-(aq)}]}{1+\mathrm{K}[\mathrm{H_3O^+(aq)}]}$$

From the solution to Exercise (38)

$$[\mathrm{HSO_3^-(aq)}][\mathrm{H_3O^+(aq)}] = k_H(\mathrm{SO_2})K_{\mathrm{a1}}p_{\mathrm{SO_2}}$$

Also, since $K = 13\ \mathrm{M}^{-1}$ and

$$\mathrm{pH} = -\log_{10}[\mathrm{H_3O^+(aq)}],$$

$$1+\mathrm{K}[\mathrm{H_3O^+(aq)}] = 1+13(10^{-\mathrm{pH}})$$

Therefore, provided pH > 2

$$1+\mathrm{K}[\mathrm{H_3O^+(aq)}] \approx 1$$

Therefore, Eq. (7.25) becomes

$$-\frac{d[\mathrm{S(IV)}]}{dt} \simeq kk_H(\mathrm{SO_2})K_{a1}p_{\mathrm{SO_2}}[\mathrm{H_2O_2(aq)}]$$
$$= kk_H(\mathrm{SO_2})K_{a1}p_{\mathrm{SO_2}}k_H(\mathrm{H_2O_2})p_{\mathrm{H_2O_2}}$$

which is independent of the pH of the solution.

40. (a) Percentage rate of decrease of $SO_2(g)$ per hour
$$= \frac{3.6\times 10^8 LR_c^* TR}{\xi_{\mathrm{SO_2}}}$$

(b) Percentage rate of decrease of $SO_2(g)$ per hour = $3.6 \times 10^{-10}\, LR_c^* Tkk_H(A)k_{\mathrm{eff}}(\mathrm{SO_2})\xi_A$

(c) 2.5% decrease of $SO_2(g)$ concentration per hour. (*Note*: This is an instantaneous upper limit, since the reaction rate R will decrease with time.)

*41.
$$\ln\frac{k_H(288.5\,\mathrm{K})}{k_H(288\,\mathrm{K})} = \frac{\Delta \mathrm{H}}{\mathrm{R}^*}\left(\frac{1}{\mathrm{T_1}} - \frac{1}{\mathrm{T_2}}\right) = \frac{-20.4\times 10^3}{8.3}\left(\frac{1}{288} - \frac{1}{288.5}\right)$$
$$= -0.0147$$

Therefore,

$$\frac{k_H(288.5\,\mathrm{K})}{k_H(288\,\mathrm{K})} = 0.985$$

and

$$k_H(288.5\,\mathrm{K}) = 0.985\{k_H(288\,\mathrm{K})\} = 4.432\times 10^{-2}\,\mathrm{mol\,L^{-1}\,atm^{-1}}$$

Therefore,

$$\Delta k_H = (4.432 - 4.5)\times 10^{-2} = -6.8\times 10^{-4}\,\mathrm{mol\,L^{-1}\,atm^{-1}}$$

Since

$$C_g = k_H\, p_g$$

when the temperature changes from 288 to 288.5 K at 1 atm

$$\Delta C_g = p_g\, \Delta k_H = -6.8\times 10^{-4}\,\mathrm{mol\,L^{-1}}$$

Therefore,

$$\text{Percentage change in } C_g = \frac{\Delta C_g}{C_g}100 = \frac{p_g \Delta k_H}{p_g k_H} = \frac{\Delta k_H}{k_H}$$
$$= \frac{-6.8\times 10^{-4}}{4.4\times 10^{-2}}\times 100 = -1.5\%$$

Hence, the percentage decrease in CO_2 in the mixed layer of the oceans is 1.5%. Since the CO_2 capacity of the atmosphere is about the same as the mixed layer of the oceans, the percentage increase in CO_2 in the atmosphere due to 0.5°C warming of oceans will be ~1.5%.

This calculation shows that the percentage increase in CO_2 in the atmosphere due to 0.5°C increase in the average temperature of the mixed layers of the world's oceans over the past 50 years is ~1.5%. However, the measured percentage increase in atmospheric CO_2 over the past 50 years is about (47 ppmv/306 ppmv)100 = 15%. Therefore, warming of oceans can account for only about 10% of the observed increase in CO_2 content of atmosphere. Therefore, it is not likely that the measured increase in atmospheric CO_2 over the past 50 years is due to warming of the oceans.

42. (a) 4 to 6 Tg(C) per year
(b) 3.0 to 5 Tg(C) per year
(c) About 70 years (see Table 2.1)

43. (a) 66 days
(b) 4 years
(c) τ_{CO} (from Table 2.1) is ~60 days. This is within margin of error of $\tau_{CO} = 66$ days, considering the uncertainties in the fluxes of CO. τ_{CO_2} (from Table 2.1) is 50–200 a! However, as noted in the footnote to Table 2.1 and in Section 8.1, the latter residence time is for atmosphere-ocean systems. The residence time for CO_2 calculated in this exercise is the average time a CO_2 molecule spends in the atmosphere.

44. 14 hours

45. (a)

$$S_2 + h\nu \xrightarrow{j_1} S + S$$

$$S + S_2 + M \xrightarrow{k_2} S_3 + M$$

$$S + S_3 \xrightarrow{k_3} S_2 + S_2$$

where $M = S + S_2 + S_3$

(b)

$$\frac{d[S]}{dt} = 2j_1 - k_2[S][S_2]M - k_3[S][S_3]$$

$$\frac{d[S_2]}{dt} = -j[S_2] - k_2[S][S_2][M] + 2k_3[S][S_3]$$

$$\frac{d[S_3]}{dt} = k_2[S][S_2][M] - k_3[S][S_3]$$

(c) At steady state

$$\frac{d[S_3]}{dt} = 0$$

Therefore,

$$\frac{[S_3]}{[M]} = \frac{k_2}{k_3}[S_2] \qquad \text{(i)}$$

With $\dfrac{d[S]}{dt} = 0$

$$2j_1 - k_2[S][S_2][M] = k_3[S][S_3]$$

Therefore,

$$\frac{2j_1}{[M]} - k_2[S][S_2] = k_3[S]\frac{[S_3]}{[M]}$$

Substituting from (i)

$$\frac{2j_1}{[M]} = k_2[S][S_2] = k_3[S]\frac{k_2}{k_3}[S_2]$$

or

$$\frac{2j_1}{[M]^2} - k_2\frac{[S]}{[M]}[S_2] = [S_2]\frac{[S]}{[M]}k_2$$

Hence,

$$\frac{[S]}{[M]}\{k_2[S_2] + [S_2]k_2\} = \frac{2j_1}{[M]^2}$$

Therefore,

$$\frac{[S]}{[M]} = \frac{2j_1}{[M]^2}\frac{1}{2k_2[S_2]} \qquad \text{(ii)}$$

But, from (i)

$$k_2[S_2] = k_3\frac{[S_3]}{[M]}$$

But, $[S_3] >> [S_1]$ and $[S_3] >> [S_2]$ and $\frac{S_3 + S_2 + S_1}{M} = 1$

Therefore,

$$\frac{[S_3]}{[M]} \simeq 1$$

and

$$k_2[S_2] = k_3 \qquad \text{(iii)}$$

Therefore, from (i) and (ii),

$$\frac{[S]}{[M]} = \frac{j_1}{k_3[M]^2}$$

(d) Since the source of the sulfur species is volcanic, M should decrease with height (above the height of the volcanoes). Hence, from

$$\frac{[S]}{[M]} = \frac{j_1}{k_3[M^2]}$$

the variation of [S]/[M] with height z should be as sketched in Figure A.7.

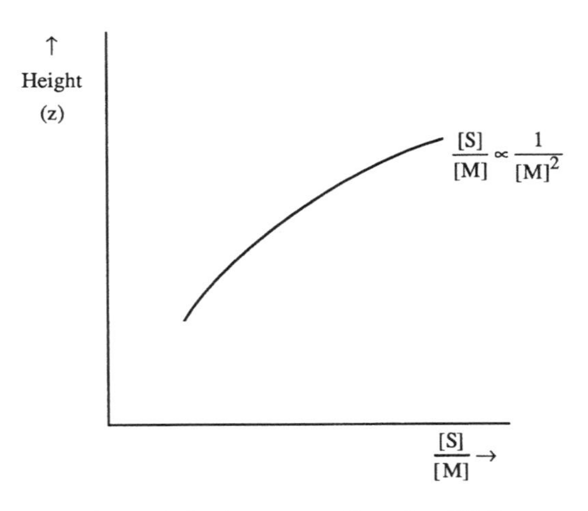

Figure A.7. Answer to Exercise 45(d).

*46. (a)

$$C_xH_y + \left(x + \frac{y}{4}\right)O_2 \rightarrow xCO_2 + \frac{y}{2}H_2O$$

From the percentage amounts of oxygen and nitrogen in air by volume (*viz.* 20.9 and 78, respectively) we have for air

$$\frac{\text{number of moles of } O_2}{\text{number of moles of } N_2} = \frac{20.9}{78} = \frac{1}{3.7}$$

Therefore, 1 mole of O_2 is associated with 3.7 moles of N_2 in air. Therefore, 3.7 $[x + (y + 4)]$ moles of N_2 will be associated with 1 mole of C_xH_y and $[x + (y + 4)]$ moles of O_2 burned.

Therefore, total number of moles of gases in the emissions for 1 mole of C_xH_y burned is

$$x \text{ moles of } CO_2$$

$$\frac{y}{2} \text{ moles of } H_2O$$

and

$$3.7\left(x + \frac{y}{4}\right) \text{ moles of } N_2$$

$$\text{Total number of moles of gas} = x + \frac{x}{y} + 3.7\left(x + \frac{y}{4}\right)$$

(b) If we include the (unreacting) nitrogen in the balanced chemical equation for *complete* combustion, we have from (a) above

$$C_xH_y + \left(x + \frac{y}{4}\right)O_2 + 3.7\left(x + \frac{y}{4}\right)N_2 \rightarrow$$
$$xCO_2 + \frac{y}{2}H_2O + 3.7\left(x + \frac{y}{4}\right)N_2$$

However, if a fraction f of fuel is provided in excess of that needed for complete combustion, and as a consequence m moles of CO_2 and n moles of CO are contained in the emissions, the chemical equation for combustion becomes

$$(1+f)C_xH_y + \left(x+\frac{y}{4}\right)O_2 + 3.7\left(x+\frac{y}{4}\right)N_2 \rightarrow$$

$$mCO_2 + nCO + (1+f)\frac{y}{2}H_2O + 3.7\left(x+\frac{y}{4}\right)N_2$$

Balancing the carbon atoms for this reaction yields

$$x(1+f) = m+n \qquad \text{(i)}$$

and, balancing the oxygen atoms, gives

$$2x+\frac{y}{2} = (1+f)\frac{y}{2} + 2m + n \qquad \text{(ii)}$$

Solving (i) and (ii) for m and n yields

$$m = x - xf - f\frac{y}{2}$$

and

$$n = f\frac{y}{2} + 2fx$$

Therefore, the mole fraction of CO in the emissions is

$$\frac{f\frac{y}{2}+2fx}{\left(x-xf-f\frac{y}{2}\right)+\left[(1+f)\frac{y}{2}\right]+\left[3.7\left(x+\frac{y}{4}\right)\right]}$$

or

$$\frac{f\left(2x+\frac{y}{2}\right)}{x(4.7+f)+\frac{y}{2}(2.85+f)}$$

(c) If the fuel is CH_2, $x = 1$ and $y = 2$. Therefore, from (b), the mole fraction of CO in the emissions is

$$\frac{3f}{7.55+2f}$$

Therefore, for $f = 0.001$ the mole fraction of unburned CO is 3.97×10^{-4} or 397 ppmv (= 0.0397%). For $f = 0.01$ it is

3.96×10^{-3} or 3,960 ppmv (= 0.396%). For $f = 0.1$ it is 3.87 $\times 10^{-2}$ or 38,700 ppmv (= 3.87%). This last concentration of CO is enough to kill you in a closed garage in about 17 minutes!

47.

$$\frac{dn_1}{dt} = 2j_a n_2 + j_c n_3 - k_b n_1 n_2 n_M - k_d n_1 n_3$$

$$\frac{dn_2}{dt} = j_c n_3 + 2k_d n_1 n_3 - j_a n_2 - k_b n_1 n_2 n_M$$

[*Hint*: See solution to Exercise 10.1. Note that there are two atoms of oxygen on the right side of Reaction (10.1) and two molecules of oxygen on the right side of Reaction (10.4).]

48. Yes, because the total numbers of oxygen atoms must be preserved.

49.
(a) $2O_3 \rightarrow 3O_2$
(b) The intermediate is O(g)
(c) Rate law for step (i) is: Rate = k_1 $[O_3]$.
Rate law for step (ii) is: Rate = k_2 $[O_3]$ $[O]$
(d) Step (ii) is rate controlling.
(e) $[O] \propto [O_3]\,[O_2]^{-1}$

50.
(a) $$\frac{d[OM]}{dt} = k_1[O][M] - k_2[OM] - k_3[O_2][OM]$$

$$\frac{d[O_3]}{dt} = k_3[O_2][OM]$$

(b) $$[OM] = \frac{k_1[O][M]}{k_2 + k_3[O_2]}$$

(c) $$k_b = \frac{k_3 k_1}{k_2 + k_3 \times \supseteq[M]}$$

See Figure A.8 for sketch of $\log k_b$ versus $\log[M]$.

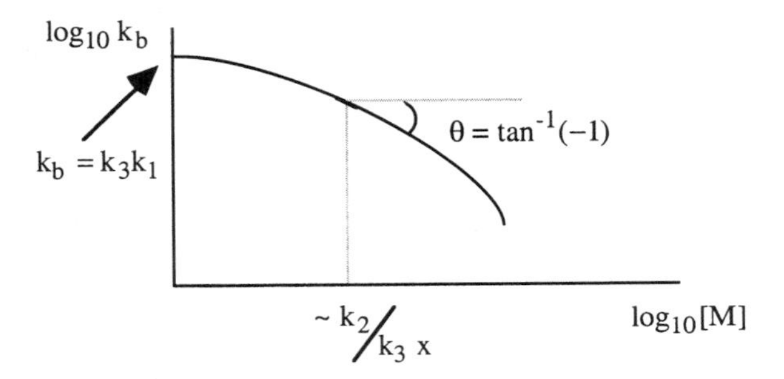

Figure A.8. Answer to Exercise 50(c).

*51. (a) $\dfrac{dX}{dt} = k_1 - k_2 X^2$ (i)

At steady state,

$$\frac{dX}{dt} = \mathrm{O}$$

Therefore,

$$X_{ss} = \left(\frac{k_1^{1/2}}{k_2}\right) = (\text{a constant})\exp\left\{\frac{1}{2}\left(\frac{300}{T} + \frac{1{,}100}{T}\right)\right\}$$

and

$$\ln X_{ss} = \text{constant} + \frac{1}{2}\left(\frac{300}{T} + \frac{1{,}100}{T}\right)$$

Hence

$$\frac{1}{X_{ss}}\frac{dX_{ss}}{dT} = \frac{d}{dT}\left(\frac{700}{T}\right)$$

$$\frac{dX_{ss}}{X_{ss}} = \frac{-700}{T^2}dT$$

For $dT = -20°\text{C}$,

$$\frac{dX_{ss}}{X_{ss}} = \frac{1{,}400}{(220)^2} = 0.029 \text{ or } 2.9\%$$

(b) Substituting $Y = X - X_{ss}$ into (i) yields

$$\frac{dY}{dt} = k_1 - k_2(Y + X_{ss})^2$$
$$= X_{ss}^2 k_2 - k_2(Y + X_{ss})^2$$
$$= -(2k_2 X_{ss} Y) + \text{terms in } Y^2 \text{ (which are small)}$$

Therefore,

$$\frac{dY}{Y} = -(2k_2 X_{ss})dt$$

and

$$[\ln Y]_{Y_0} = -2k_2 X_{ss} \int_0^t dt$$
$$\frac{\ln Y}{Y_0} = -2k_2 X_{ss} t$$

or

$$Y = Y_0 \exp(-2k_2 X_{ss} t)$$

The relaxation time (τ) is the time for y to decline to exp(−1) of an initial value Y_0. Therefore, from the last expression

$$\tau = \frac{1}{2k_2 X_{ss}} = \frac{1}{2\left\{10\exp\left(\frac{-1,100}{200}\right)(5 \times 10^{-7})\right\}} = 1.48 \times 10^7 \text{ s}$$

$= 172$ days

52. (a) $H + O_3 \rightarrow HO + O_2$

$HO + O \rightarrow H + O_2$

Net: $O + O_3 \rightarrow 2O_2$

(b) $OH + O_3 \rightarrow HO_2 + O_2$

$HO_2 + O \rightarrow OH + O_2$

Net: $O + O_3 \rightarrow 2O_2$

53. (a) If (iia) dominates the net effect is

$$2O_3 + h\nu \rightarrow 3O_2$$

If (iib) dominates, there is no net effect (ClO never gets to Cl)

(b) $$\frac{d[O_3]}{dt} = -k_4[Cl][O_3]$$

$$\frac{d[Cl]}{dt} = j_2[(ClO)_2] + k_3[ClOO][M] - k_4[Cl][O_3]$$

$$\frac{d[(ClO)_2]}{dt} = k_1[ClO]^2[[M] - j_2[ClO_2]$$

$$\frac{d[ClOO]}{dt} = j_2[(ClO)_2] - k_3[ClOO][M]$$

(c) $$[Cl] = \frac{2k_1[M][ClO]^2}{k_4[O_3]}$$

(d) $$\frac{d[O_3]}{dt} = -2k_1[M][ClO]^2$$

(e) $[O_3] \propto t^3$

*54. (a) $$\frac{d[O^*]}{dt} = j_1[O_3] - k_1[O^*][M] - k_2[O^*][H_2O]$$
$$= 0 \text{ (at steady state)}$$

Therefore,

$$[O^*] = \frac{j_1[O_3]}{k_1[M] + k_2[H_2O]}$$
$$\simeq \frac{j_1[O_3]}{k_1[M]}$$

Hence, the molecular fraction of O*, say f (O*), is

$$f(O^*) \simeq \frac{j_1 f(O_3)}{k_1[M]} = \frac{10^{-4}(2 \times 10^{-6})}{10^{-11}(5 \times 10^{17})} = 4 \times 10^{-17}$$

Also,

$$[HO] + [HO_2] = [\text{odd hydrogen}]$$

Therefore,

$$\frac{d[\text{odd hydrogen}]}{dt} = 2k_2[O^*][H_2O] - 2k_6[HO][HO_2] \quad \text{(i)}$$

If the k_4 and k_5 steps occur often compared with the steps associated with k_2 and k_6, they determine the concentrations of HO and HO_2. Therefore, at steady state

$$k_4[HO][O_3] \simeq k_5[HO_2][O_3] \qquad \text{(ii)}$$

or

$$\frac{[HO_2]}{[HO]} \simeq \frac{k_4}{k_5} \simeq \frac{2\times10^{-14}}{3\times10^{-16}} \simeq 70 \qquad \text{(iii)}$$

From (i) and (iii), and with d[odd hydrogen]/dt = 0,

$$k_2[O^*][H_2O] - k_6\frac{k_5}{k_4}[HO_2]^2 = 0$$

Therefore,

$$[f(HO_2)]^2 = \frac{k_2k_4}{k_5k_6} f(O^*)f(H_2O)$$

$$f(HO_2) = \left\{\frac{(2\times10^{-6})(2\times10^{-14})}{(3\times10^{-16})(3\times10^{-11})}(4\times10^{-17})(2\times10^{-6})\right\}^{1/2}$$
$$\simeq 2\times10^{-10}$$

Using (ii),

$$f(HO) = \frac{f(HO_2)}{70} \simeq 3\times10^{-10}$$

(b) The mean lifetime of odd hydrogen is

$$\tau_{\underset{\text{hyd}}{\text{odd}}} = \left\{\frac{\text{loss rate of odd hydrogen}}{\text{concentration of odd hydrogen}}\right\}^{-1}$$

Or, since $[HO_2] = 70\,[HO]$

$$\tau_{\underset{\text{hyd}}{\text{odd}}} = \left\{\frac{2k_6[HO][HO_2]}{[HO_2]}\right\}^{-1}$$

Therefore,

$$\tau_{\underset{\text{hyd}}{\text{odd}}} = \{2k_6[HO]\}^{-1}$$
$$= \{2k_6 f(HO)[M]\}^{-1}$$
$$= \{2(3\times10^{-11})(3\times10^{-10})(5\times10^{17})\}^{-1} = 3{,}509\ \text{s}$$

$$\tau_{\text{odd}} \simeq 1\ \text{h}$$

(c)

$$\text{Let } N = \frac{\text{number of O}_3\text{ molecules destroyed per second}}{\text{number of odd hydrogen atoms produced per second}}$$

$$\simeq \frac{k_4[\mathrm{HO}][\mathrm{O}_3] + k_5[\mathrm{HO}_2][\mathrm{O}_3]}{\dfrac{[\mathrm{HO}]+[\mathrm{HO}_2]}{\tau_{\mathrm{odd}}}}$$

Or, using (ii) and (iii),

$$\begin{aligned} \mathrm{N} &\simeq 2k_5[\mathrm{O}_3]\tau_{\mathrm{odd}} = 2k_5 f(\mathrm{O}_3)[\mathrm{M}]\tau_{\mathrm{odd}} \\ &= 2(3\times10^{-16})(2\times10^{-6})(5\times10^{17})(3{,}509) \\ &= 2 \end{aligned}$$

Appendix III

Atomic weights[a]

Element	Symbol	Atomic Weight
Actinium	Ac	(227)
Aluminum	Al	26.9815
Americium	Am	(243)
Antimony	Sb	121.75
Argon	Ar	39.948
Arsenic	As	74.9216
Astatine	At	(210)
Barium	Ba	137.34
Berkelium	Bk	(247)
Beryllium	Be	9.0122
Bismuth	Bi	208.980
Boron	B	10.811
Bromine	Br	79.904
Cadmium	Cd	112.41
Calcium	Ca	40.08
Californium	Cf	(251)
Carbon	C	12.01115
Cerium	Ce	140.12
Cesium	Cs	132.905
Chlorine	Cl	35.453
Chromium	Cr	51.996
Cobalt	Co	58.9332
Copper	Cu	63.546
Curium	Cm	(247)
Dysprosium	Dy	162.50
Einsteinium	Es	(252)
Erbium	Er	167.26
Europium	Eu	151.96
Fermium	Fm	(257)
Fluorine	F	18.9984

(cont.)

Element	Symbol	Atomic Weight
Francium	Fr	(223)
Gadolinium	Gd	157.25
Gallium	Ga	69.72
Germanium	Ge	72.59
Gold	Au	196.967
Hafnium	Hf	178.49
Helium	He	4.0026
Holmium	Ho	164.930
Hydrogen	H	1.00794
Indium	In	114.82
Iodine	I	126.904
Iridium	Ir	192.2
Iron	Fe	55.847
Krypton	Kr	83.80
Lanthanum	La	138.91
Lawrencium	Lw	(260)
Lead	Pb	207.19
Lithium	Li	6.939
Lutetium	Lu	174.97
Magnesium	Mg	24.305
Manganese	Mn	54.938
Mendelevium	Md	(258)
Mercury	Hg	200.59
Molybdenum	Mo	95.94
Neodymium	Nd	144.24
Neon	Ne	20.179
Neptunium	Np	(237)
Nickel	Ni	58.69
Niobium	Nb	92.906
Nitrogen	N	14.0067
Nobelium	No	(259)
Osmium	Os	190.2
Oxygen	O	15.9994
Palladium	Pd	106.4
Phosphorus	P	30.9738
Platinum	Pt	195.09
Plutonium	Pu	(244)
Polonium	Po	(209)
Potassium	K	39.102
Praseodymium	Pr	140.907
Promethium	Pm	(145)
Protactinium	Pa	(231)
Radium	Ra	(226)
Radon	Rn	(222)

(cont.)

Element	Symbol	Atomic Weight
Rhenium	Re	186.2
Rhodium	Rh	102.905
Rubidium	Rb	85.47
Rutherfordium	Rf	(261)
Ruthenium	Ru	101.07
Samarium	Sm	150.35
Scandium	Sc	44.956
Selenium	Se	78.96
Silicon	Si	28.086
Silver	Ag	107.868
Sodium	Na	22.9898
Strontium	Sr	87.62
Sulfur	S	32.066
Tantalum	Ta	180.948
Technetium	Tc	(98)
Tellurium	Te	127.60
Terbium	Tb	158.924
Thallium	Tl	204.38
Thorium	Th	232.038
Thulium	Tm	168.934
Tin	Sn	118.71
Titanium	Ti	47.88
Tungsten	W	183.85
Uranium	U	238.03
Vanadium	V	50.942
Xenon	Xe	131.29
Ytterbium	Yb	173.04
Yttrium	Y	88.905
Zinc	Zn	65.39
Zirconium	Zr	91.22

[a] Based on an atomic weight for carbon-12 of 12.000. Values in parentheses are for the most stable known isotopes.

Appendix IV

The international system of units (SI)[a]

Quantity	Name of Unit	Symbol	Definition
		Basic units	
Length	meter	m	
Mass	kilogram	kg	
Time	second	s	
Electrical current	ampere	A	
Temperature	degree Kelvin	K	
		Derived units	
Force	newton	N	$kg\,m\,s^{-2}$
Pressure	pascal	Pa	$N\,m^{-2} = kg\,m^{-1}\,s^{-2}$
Energy	joule	J	$kg\,m^2\,s^{-2}$
Power	watt	W	$J\,s^{-1} = kg\,m^2\,s^{-3}$
Electric potential difference	volt	V	$W\,A^{-1} = kg\,m^2\,s^{-3}\,A^{-1}$
Electrical charge	coulomb	C	$A\,s$
Electrical resistance	ohm	Ω	$V\,A^{-1} = kg\,m^2\,s^{-3}\,A^{-2}$
Electrical capacitance	farad	F	$A\,S\,V^{-1} = kg^{-1}\,m^2\,s^4\,A^2$
Frequency	hertz	Hz	s^{-1}
Celsius temperature	degree Celsius	°C	K – 273.15
Temperature interval	degree	degree or °	K or °C need not be specified

[a] SI units are the internationally accepted form of the metric system. SI units are being used increasingly in chemistry, as in other sciences, but some older units persist. This book reflects this dichotomy, although SI units have been used as much as possible. Some useful conversion factors between various units are given in Appendix V.

Prefixes used to construct decimal multiples of units

Multiple	Prefix	Symbol	Multiple	Prefix	Symbol
10^{-1}	deci	d	10	deca	da
10^{-2}	centi	c	10^{2}	hecto	h
10^{-3}	milli	m	10^{3}	kilo	k
10^{-6}	micro	μ	10^{6}	mega	M
10^{-9}	nano	n	10^{9}	giga	G
10^{-12}	pico	p	10^{12}	tera	T
10^{-15}	femto	f	10^{15}	peta	P
10^{-18}	atto	a	10^{18}	exa	E

Appendix V
Some useful numerical values

Universal constants	
Universal gas constant, in SI units (R^*)	$8.3143\,J\,deg^{-1}\,mol^{-1}$
Universal gas constant, in "chemical units" (R_c^*)	$0.0821\,L\,atm\,deg^{-1}\,mol^{-1}$
Boltzman's constant (k)	$1.381 \times 10^{-23}\,J\,deg^{-1}$ molecule^{-1}
Avogadro's number (N_A)	6.022×10^{23} molecules mol^{-1}
Faraday constant (F)	96,489 C equiv^{-1}
Planck's constant (h)	$6.6262 \times 10^{-34}\,J\,s$
Velocity of light in a vacuum (c)	$2.998 \times 10^{8}\,m\,s^{-1}$
Other values	
Number of molecules in $1\,m^3$ of a gas at 1 atm and 0°C (Loschmidt's number)	2.69×10^{25} molecules m^{-3}
Volume of 1 mole of an ideal gas at 0°C and 1 atm	22.415 liters
Ion-product constant for water at 25°C and 1 atm (K_w)	$1.00 \times 10^{-14}\,mol^2\,liter^{-2}$

Conversion factors

1 bar = 10^5 Pa

1 atm = 1.013 bar = 1,013 mb = 760.0 Torr

For a second-order rate coefficient:

$1\,cm^3\,s^{-1}$ molecule^{-1} = 6.02×10^{20} liter $s^{-1}\,mol^{-1}$

For a third-order rate coefficient:

$1\,cm^6\,s^{-1}$ molecule^{-2} = 3.63×10^{41} liter^2 $s^{-1}\,mol^{-2}$

1 kcal mol^{-1} = 4.18 kJ mol^{-1}

1 eV = 96.489 kJ mol^{-1}

$\ln x = 2.3026 \log x$

Appendix VI
Suggestions for further reading

The following suggestions for further general readings in atmospheric chemistry are listed in approximately increasing order of difficulty and/or comprehensiveness.

Hobbs, P. V., *Basic Physical Chemistry for the Atmospheric Sciences*, 2nd edition, Cambridge University Press, New York, 2000.

Brimblecombe, P., *Air Composition and Chemistry*, Cambridge University Press, Cambridge, 1996.

Mészáros, E., *Atmospheric Chemistry: Fundamental Aspects*, Elsevier Scientific Publishing Co., Amsterdam, 1981.

Harrison, R. M., S. J. de Mora, S. Rapsomanikis, and W. R. Johnston, *Introductory Chemistry for the Environmental Sciences*, Cambridge University Press, Cambridge, 1991.

Schlesinger, W. H., *Biogeochemistry: An Analysis of Global Change*, Academic Press, New York, 1997.

Wayne, R. P., *Chemistry of Atmospheres*, Oxford University Press, Oxford, 1991.

Warneck, P., *Chemistry of the Natural Atmosphere*, Academic Press, 1988.

Finlayson-Pitts, B. J., and J. N. Pitts, Jr., *Atmospheric Chemistry*, John Wiley and Sons, Inc., New York, 1986.

Seinfeld, J. H., and S. N. Pandis, *Atmospheric Chemistry and Physics*, John Wiley and Sons, Inc., New York, 1998.

Index

Italic numbers refer to figures or tables.

absorption bands, 40–43, *41*, 62
absorption coefficient, 37–38
absorption cross-sections, 58, *60*
absorption mass efficiency, 38
absorption spectrum, 40–41, *41*
absorptivity, 37, 40, 53
accumulation mode, *86*, 86–87, 89, 96, 98, 102
acetic acid (CH_3COOH), *122*
acetyl radical (CH_3CO), 159
acid deposition, 100, 162
acid dissociation (*or* equilibrium) constant, 70, 123, 126
acidic aerosols, 100, 149
acid rain, 79, *112*, 131–136, *134–137*
acidic solutions, *136*
acids. *See under* individual names
actinic flux, 57–58, *60*, *139*, 139–140, 191–192, *192*
activated droplets, 116
aerobic
  bacteria, 64, 149
  conditions, 149
aeronomy, 32
aerosols, 82–110
  absorption of em radiation, 37–40
  accumulation mode, *86*, 87–89, 96, 98, 102
  acidic, 100, 149, 194
  Aitken. *See* Aitken nuclei
  altitude dependence, 83, *84–85*, *179*
  aluminum in, 99
  anthropogenic sources, 50–51, 57, 94–96, *96*, *103*, *105*, 119, 153, 156, 179–180
  carbonaceous, 51, 95–96, 99
  cloud interstitial, 112, 140
  coagulation, 87, 100–102, 112
  coarse particle mode, *86*, 86–87, 89
  collection by hydrometeors, 131–133, *132–133*
  collisions, 97, 100–102, *101*, 131–133, *132–133*, 196
  composition, 82, 90, 95, 97–99, 160
  concentrations
    around clouds, 137, *138–139*, *141*
    background, 83, *84–85*
    in stratosphere, 179, *179*
    in troposphere, 82–92, *84*, 140, 196, 200–201
  continental, 83, *84–85*, 87, *88–89*, 92–93, 97–99, *98*, *103*, 104, *106*, 118–119
  definition, 82
  deliquescent point, 90
  deposition velocity, 100–102, *101*, 193
  direct radiative forcing, 57–*58*
  effects, 50–51, 57, 104, 151, 156, 160–163, 180–182
  enrichment, 95, 99

from evaporating clouds, 137, *139*, *141*
extinction of em radiation, 37–40
formation, 91–97, *96*, 133–140, *139*, *141*, 153, 155, 179–183, *182*, 193
geographical distribution, *105–106*, 104–110
geometric mean diameter, 90–91
geometric standard deviation of, 90–91
giant, *86*, 87–88, *93*, 93–94, 97, *103*, *110*
homogeneous bimolecular nucleation, 137
impaction, 100–102, *103*
indirect radiative effects on clouds, 57–*58*
inorganic, *98*, 98–99
in situ formation, 94
interstitial, 112, 140
Junge distribution, 85–88, *86*
large, *86*, 86–87, 92–94, 98, 100, *103*, *110*, 133, *133*, 156
lead, 162
log-normal distribution, *90*, 90–91
marine, 83, *84–85*, 88, *88*, 93–95, 98–99, *103*, 104, *105*, 118–119, 137–140, *138–139*, *141*, 194, 200
mass concentration, 82–85, *85*, 98–99, 102, 186
median diameter, 91
mineral dusts, 92, *106*
mixed, 118
modeling of, 91, *106*
modification by clouds, 111–112, *112*, *120*, 137–140, *139*, *141*
natural sources, 68, 88, 91–97, *93*, *96*, *103*, *105–106*, 179–183, *181–182*
nitrates, *24*, 64, 90, 95–96, *96*, 98–99, 136, *148*, 149, 155, 159, 163
normal distribution, *90*, 90–91
nucleation scavenging, 111–116, *114*, 125, 134–135
nucleus (or nucleation) mode, *86*, 86–87
number distribution. *See* aerosol: size distribution,
optical depth (*or* thickness), 39–42, 45, 50, 54, 104–105, *105*, 180–*181*, 191, 196
organic, 50–51, 64, 91–99, *96*, 119
in polluted air, 50, 83, *86*, *89*, 89, 94–96, *103*, 134–135, 156, 160–163
precipitation scavenging, 102–103, *103*, 113, 131–133, *135*, *139*
production. *See* formation and sources
radiative forcing, 56–57, *58*
reactions on surfaces, 78, 111
residence time, 15–20, *18–19*, 56, 93, 100, 102–104, *103*, 151, 171, 187, 193–194, 200
scattering of em radiation, 37–40
scavenging, 80, 100–102, *101*, 111–116, 131–140, *132*, *134–135*, *139*
sedimentation, 97, 100–102, *101*, 196
silver, 99, 163
sinks, 100–102, *103*
size distribution (*or* spectra), 82–92, *86*, *88–90*, *103*
size distribution function, 83
soluble, 90, 97–98, *98*, 111, *114*, 114–118
sources, *88*, 91–97, *93*, *96*, *103*, 103–104, 118–119, *120*, *138–139*, 140–141, *141*, *148*, 149–153, *150*, 156–157, *157*, 182–183, 193
spatial scales of variability, *19*
in stratosphere, 99–100, 179–183
sulfates. *See* sulfates
surface area distribution, *89*, 89–90, *103*, 195
terminal settling (*or* fall) velocity, 100, *101*, 102
total number, 82–91
transformation, 95–97, *139*, 139–140
transport, 99, 104–105, *105*, *139*–140, 153, 162, 182
visibility reduction due to, 160–162
volume distribution, *89*, 89–90, 195
volume distribution function, 83–90, 92, 195–196

aerosols (*cont.*)
water-soluble, 90, 97–98, *98*, 113, 116, 118
wettable, 113, 118
aerosol cans, 171
aerosol-free air
extinction, 160
meteorological range, 160–161
scattering, 37
aerosol optical depth. *See* optical depth
aerosol particles. *See* aerosols
air
apparent molecular weight, 154, 207
clean, *24*, 37, 40, 74, 100, 134, *134–135*, 155, 160–162
composition, 1, 15, 21–32, *24*, 62, 134, *134–135*
continental, 83, *84–85*, 87, *88–89*, 91–94, 98–100, *103*, 103–104, *106*, 118–119, 186
dynamic viscosity, 97, 100, 196
marine, 83, *84–85*, 87–*88*, 94, 98–99, 118, 194
mean free path, 26–29, *28*, 97, 100
polluted, 37, 153–163
air bubble bursting, 92–94, *93*
aircraft effluents, 164
air/fuel mixture, 154
air-fuel ratio, 153–154
airglow, *27*, 29, 32
air pollution, 38, 110, *110*, 153–163, 187
abatement laws, 156
effect on visibility, 160–162
effects, 104–110, 156–163
low temperature sources, 155
Aitken, John, 110
Aitken (*or* condensation) nuclei. *See also* aerosols
concentrations (*or* count), 82–91, *84–86*, *88*, 94–95, 137, *138–139*, *141*, 179
sea salt contribution, 88, *88*
sinks, 100–102, *103*
sources, 91–96, 102–103, *103*, *138–139*, 140–141, *141*, 180
in stratosphere, *179*, 179–183, *182*
supply to upper atmosphere, 104, 140–141, *141*
Aitken (*or* condensation) nucleus counter, 82
Alaid, *181*
albedo (*or* reflectivity), 54–55, 142, 197, *217*
aldehydes, 158–159
algae, *11*, 81, 91
alkaline solutions, *136*
alkanes, 99
alkenes, 99, 159
aluminum (Al), *2*, 95, 99, 235
Amazon Basin, 57
amino acids, 3–4
ammonia (NH_3), 1–3, 10, 18, *18*, *24*, 64–65, 98, 124, *148*, 148–149, 190
ammonium (NH_4^+), *18*, 98–99, 136–137, *137*, *148*, 194
ammonium chloride (NH_4Cl), *24*
ammonium nitrate (NH_4NO_3), *24*, 90, 162
ammonium salts, 149
ammonium sulfate [$(NH_4)_2SO_4$], 10, *24*, 93, 98, 198
ampere, 238
anaerobic conditions, 149
animals, *11*, 14, 64–65, *144*–145, *148*
Antarctic
aerosol, 180–181, *181*
optical depths, 180–181, *181*
ozone hole, 79, 171–179, *173–175*
polar stratospheric clouds (PSC), 172–177, 180–181, *181*, 183–184
stratosphere, *166*, 171–184, *173–175*, *179*, *181–182*
vortex, 172–176, *175*
anthropogenic
aerosols, 50–51, 57, 83–85, 94–96, *96*, *98*, *103*, 104–110, *105*, 118–119, 151–153, 156, 160–163, 179–180
air pollution, 38, 104–110, *110*, 153–163, 187
gases, 64–65, 95–97, 119, *144*, 145–

163, *148*, *150*, *157*
radiative forcing, 56–57, *58*
apparent molecular weight of dry air, 154
aquatic organisms, 143
aqueous-phase chemical reactions, 112, *112*, 119–*120*, 125–137, *127*, *134–135*, *139*
Arctic
air pollution, 162
optical depths, *181*
ozone hole, 177–178
argon (Ar), 2, *2*, 23–24, *24*, 61, 68, 235
Asian aerosols, 104
assimilation. *See* photosynthesis
atmosphere
Earth. *See* Earth's atmosphere
isothermal, 43–44, *46*, 53, 191
Jupiter, 1, 201
Mars, 1, 8, 185
Neptune, 1
Saturn, 1
unit of pressure, 240
Uranus, 1
Venus, 1, 8, 12, 185
atmosphere-ocean system, 6
atmospheric
chemistry. *See* specific entries
electricity, 82, 104
radiation, 5–6, 11, 30–31, 33–62, 82, 104, 119–121, 165, 171, 178, 183, 186, 196–198
windows, 34, 41
atomic hydrogen (H), 1–2, 4, 6, *24*, 26–27, *27*, 29–30, 74–76, 159, 169–170, 184, 188–189, 193, 203–206, 231, 236
atomic oxygen (O, O*), 3–4, 6, *24*, *27*, 29–30, 61, 72–75, *73*, 77, *77*, 155, 158, 167–172, 176, 180, 182–184, 188–189, 193, 195, 202–206, 229–233, 236
atomic weights, 235–237
atoms, vibrational energy, 43
atto-, 239
Augustine, *181*
aurora, *27*, 32
automobiles, 147, 153–158, *157*
Avogadro's number, 240

bacteria
from bubble bursting, 93–94
degradation, 64, 145, 189–190
methanogenic, 189, 206
in oceans, 6
in plants, 91, 147–151, *148*
role in fixation, *148*, 148–149
in soils, 65, *148*
bar, 240
Beer, August, 61
Beer's law, 39
bicarbonate ion (HCO_3^-), 5, 8, 69–71, 186
bimolecular
nucleation, 137, 180
reaction, 190, 209
biogenic
reactions, 145, 151
sources, 63–65, 91, *144*, *148*, *150*
biogeochemical cycling. *See* chemical cycles
biogeochemistry, 241
biological
effects on atmosphere, 10
processes in oceans, 64, 71
sources of gases, 63–68
biomass burning
aerosols, 50, 57, 83, 92, 94, 97, *144*, *148*, *150*, 152–153, *157*, 163
in chemical cycles, 143–152, *144*, *148*, *150*
elemental carbon, 92
gases, 64–65, *144*, 147, *148*, 148–149, *150*, 153, *157*, 163
radiative forcing by, 56–57, *58*
biosphere, 1–2, 7, *7*, 10, 72, 96, 149, 63–64, 143–145
bisulfite ion (HSO_3^-), 125–130, 198, 221–222
blackbody, 35–36, 53–54
blackbody radiance, 53
blackbody spectra, *36*
black carbon (*or* elemental *or* graphitic carbon), 38, 51, 92

Boltzmann's constant, 240
Boltzmann, L., 32
boundary layer, *19*, 79–81, 138–140, 194, 200
bromine (Br), 99, 235
bromine oxide (BrO), 170, 184
Brownian
  coagulation, 102
  diffusion, 138
  motion, 97, 101–102
bubble bursting
  film droplets, 93, *93*
  jet drops, 93, *93*
  materials from, *93*, 93–94, 99, 119, 196, 215
butane (C_4H_{10}), 158

calcium (Ca), *2*, 235
calcium carbonate (*or* calcite) ($CaCO_3$), 5, 8, 69–71, 192
calcium ion (Ca^{2+}), 8, 69, 99, *137*, 192, 208
calcium sulfate ($CaSO_4$), 93
Cambrian, *11*
carbohydrates, 5
carbon (C)
  atomic weight, 235
  biogeochemical cycle, 143–147, *144*
  black (*or* elemental *or* graphitic), 38, 51, 92, 162
  budget, 6, 143–147, *144*
  on Earth, 1–12, *2*, 38, 51, 63–72, 92, 143–147, *144*, 190
  elemental (*or* black *or* graphitic), 38, 51, 92, 162
  graphitic (*or* black *or* elemental), 38, 51, 92, 162
  mass entering atmosphere, 207
  near Earth's surface, *7*
  organic, 5–7, 14–15, 51, 64, 92, *144*, 144–145, 185
  in rocks, 12, 69
carbon-12, 14, 207, 237
carbon-14, 14–15, 185
carbon dioxide (CO_2)
  absorption bands, *42*, 43–44, 48
  atmospheric concentration, *9*, *11*, 21–23, *24*, 62, 147, 185
  atmospheric reservoir, *7*, 72, *144*, 144–145, 147
  in carbon cycle, 143–147, *144*
  change with time, *9*, *11*, 163
  dissolution in ocean, 9, *144*, 144–145, 199
  dissolution in water, 8, *122*, 123–124, *125*
  effect on stratospheric temperatures, 203
  evolution in Earth's atmosphere, 6–12, *9*, *11*
  fluxes from anthropogenic sources, *144*, 147
  global warming by, 10, 56–57, 147, 163
  greenhouse effect, 10, 56–57, 147, 163
  Henry's law constant, *122*, 123–124, *125*
  on Mars, 1, 8
  photolysis, 57, 65, 158, 189, 191, 204
  photosynthesis, role in, 5–7, 14, 63–64, *144*, 144–145, 147
  radiative forcing by, 56, *58*, 62
  residence time, *18*, 56, 143–145, 147, 200, 224
  respiration, 7, 63–64, *144*, 147
  sinks, 7, 72, *144*, 147
  solubility in water, 9, 198, *See also* Henry's law constant
  sources, 5, 63–65, 69, 72–73, *73*, *144*, 147, 153, 159, 163, 193
  weathering of calcite by, 69–71
  on Venus, 1, 8, 12
carbon-containing gases, principal sources of, 143–147, *144*
carbon disulphide (CS_2), *24*, 64, 68, 95, *150*, 150–151, 187–188, 206
carbon monoxide (CO), 194
  from automobiles, 154–155, *157*
  chemical cycle, 143–147, *144*
  in clean air, 155
  concentration, *24*, 155, 202, 229
  deposition velocity, 80
  emission factors, 156–157, *157*

in polluted air, 154–155, 158–159, 163
in primitive air, 2–3, 10
reactivity, 23, 154
residence (*or* lifetime) time, *18–19*, 194, 200, 209–210, 224
sinks, *73*, 74–75, *144*, 147, 194
sources, *24*, 64–65, 68, 73, *73*, 76, *144*, 147, 153–155, 163, 182, 193
spatial variability, *19*
toxicity, 154–155
carbonaceous aerosol, 51, 95–96, 99
carbonates, *7*–8, 12, 69–71
carbonic acid (H_2CO_3), 8, 70–71, 81, 123, 126, 135, 192, 198, 208
carboniferous, *11*
carbonyl sulfide (COS), *18*, *24*, 64–68, 95, *150*, 150–151, 182, *182*, 187, 206
carcinogens, 99
cars. *See* automobiles
catalyst
for automobiles, 154–155
in stratosphere, 169–177, 188, 203–204
catalytic cycles
in Earth's early atmosphere, 4
in the stratosphere, 169–172, 176–177, 188, 203–204
centi-, 239
cesium (Cs), 61, 235
CFC-11 ($CFCl_3$), 171, 176, 184
CFC-12 (CF_2Cl_2), 171, 176, 184
Chapman Reactions, *59*, 167–169, 184, 187–188
Chapman, S., 167, 184
Chappuis band, 62
chemical cycles, 4, 143–152, 169–172, 176–177, 188, 203–204
chemical families, 77–78, *96*
chemical reaction cycles, 78
chemical reactions in aqueous solutions, 112, *112*, 125–128, 132–137, *134–135*, 142
chemical species
spatial variability, *19*, 19–20
temporal variability, 20
chemical symbols, 11, *19*, *24*, *235*
chlorine (Cl)
atomic weight, 235
from chlorofluorocarbons, 171, 176–178, 184, 204
in hydrosphere, *2*
and stratospheric ozone hole, 171–179, 184, 203–204
chlorine (Cl_2), 176
chlorine dioxide (OClO or ClOO), 176–178, 184, 203, 232
chlorine ion (Cl^-), 99, *137*
chlorine nitrate ($ClONO_2$), 174–177
chlorine oxide (ClO), 170, 172–178, 184, 203–204, 231
chlorine oxide dimer [$(ClO)_2$], 176–177, 203, 232
chlorofluorocarbons (CFC)
elimination of, 178
as greenhouse gases, 56, 163, 184
lifetime, 171, 178
photolysis, *59*, 171
radiative forcing by, 56, *58*
replacements for, 184
residence times, *18–19*, 171
sources, 155, 163, 171
spatial variability, *19*
and stratospheric ozone hole, 163, 171–177, 178–179, 184
chloroform ($CHCl_3$), 65
cigarette smoke, 155
classical (*or* London) smog, 156
climate, 10, 11, 62, 96
cloud chemistry, 79, 111–142
cloud condensation nuclei
chemical composition, 112, 119–120, *120*, 137
concentrations, 118–119
effect of particle size on, 118
effect of particle solubility on, 118
effects on reflectivity of clouds, 57–58, *58*, 142, 196–198
in formation of cloud droplets, 111–121, *112*, *114*, *120*
mixed, 118
nucleation scavenging, 111–113, *112*, 134–135

cloud condensation nuclei (*cont.*)
- production rate, 197
- residence time, 197
- sinks, *120*
- sources, 118–119, *120*, 140

cloud droplets
- chemical reactions in, 10, 78, 94, 111–112, *112*, 119, 123, 125–137, *135*, 142
- collisions, 142
- dissolution of gases in, *112*, 112–113, 121–126, *122*, *125*
- evaporation, 94, 101–102, 138–139, *139*, *141*
- growth, 111–117, *112*, *114*, *120*
- insoluble components in, 111
- nucleation, 111–121, 137
- pH, 124, 126–128, *134*, 135–136
- soluble components in, 111–119, *112*, *114*, *120*
- supersaturation over, 113–119, *114*

cloud interstitial aerosol, 112, *138*, 140

clouds
- aerosol production by, 104, *135*, 137–140, *138–139*, *141*
- aerosol removal mechanisms in, *132–133*, *135*, 138–139, *139*
- albedo, 197, *217*
- chemical reactions in, 10, 78, 94, 111–113, *112*, 119–120, *120*, *122*, 122–123, *125*, 125–134, *127*, *134–135*, 140, 142
- collision and coalescence of droplets in, 112, *112*
- continental, 83, 87, 94, 100, 104, 119, 186
- convective, 33, 119–120, *120*, *138–139*, 138–140, *141*
- detrainment of air from, 119–120, *120*, *139*, 139–140, *141*
- dissolution of gases in, 113, 121–125, 131
- effects of aerosols on, 57, 104, 111–121, *112*, *114*, *120*
- entrainment of air into, *139*, *141*, 141–142
- humidity around, 137, *138–139*
- liquid water contents, 119, 129–131, *134*, 198, 220
- longwave radiation from, 33–34, *34*
- marine, 95, 98–99, 119–121, *120*, 137–140, *138–139*
- reflectivity, 57, 119–121, 142, 196–198
- scavenging by, 80, 112–121, 131–133, *139*, *141*
- stratiform, 119–121, *120*, 140
- supersaturations in, 111–118, *114*

coagulation of aerosols, 97, 100–102, *103*

coal. *See* combustion

coalescence efficiencies, 133, *133*

coal-fired power plants, *157*

coal mines, *144*

coarse particle mode, *86*, 87, 89

collection efficiencies, *132*, 132–133, *134*

collisional scavenging, *133*, 140

column optical depth. *See* optical depth

combustion
- in automobiles, 147, 153–158, *157*, 201
- coal, 65, 94, 151–156, *157*
- complete, 153–154, 201–202, 227
- of conventional fuels, *144*, *148*, *150*, 153–156, *157*
- of hydrocarbons, 64, 153–156, 201–202
- as source of chemicals, 65, 99, *144*, *148*, *150*, 153–156, *157*

comets, 4

condensation
- gases, 57, 87, 95–98, 138–142, *139*, *141*, 180. *See also* gas-to-particle conversion
- water, 82, 101, 111–121, 142, 174

condensation nucleus concentrations (*or* count). *See* Aitken nuclei

condensation (*or* Aitken) nucleus counter, 82

continental air, 83, *84–85*, 87, *88–89*, 91–94, *103*, 103–104, *106*, 118,

186–187
continental clouds, 83, 87, 94, 100, 104, 119, 186
copper (Cu), 99, 155, 235
copper smelters, *157*
cosmic rays, 14
cosmos, 2
coulomb, 238
Cretaceous, *11*
crust. *See* Earth's crust
Crutzen, P., 184
cryosphere, 143
cumulus clouds. *See* clouds, convective

deca-, 239
deci-, 239
deforestation, *144*
dehydration in stratosphere, 174, 178
deliquescent point, 90
denitrification, *148*, 148–149, 172, 178
density-weighted path length, 48
deposition
  dry, 80, 100–102, *101*, *148*, 149–151, *150*
  wet, 80, 100, 102, *148*, 149–151, *150*
deposition velocity
  of aerosols, *101*, 101–102
  of a gas, 80
deserts, *85*, 92, 99, *106*
detrainment from clouds, *120*, *139*, 139–140, *141*
detrious decomposition, *144*
diffuse radiation, 42, 52
diffusional collisions, 132–133, *133*
diffusiophoresis, 101, 140, 187
dimethyldisulfide (CH_3SSCH_3), *24*, 64
dimethylsulfide (CH_3SCH_3 *or* DMS)
  amount in atmosphere, *150*
  concentration, *24*, 151, 194–195, 211–213, *214*
  effects on clouds, *120*, 187
  from oceans, 64, 81, 119–121, *120*, 138, *150*, 150–151, 194–195, 211–213
  oxidation state of sulphur in, 187, 206
  residence time, *18–19*, 151, 194
  solubility, *122*, 140
  as source of sulfates, 95–96, *96*, 119–120, *120*, *150*
  in sulfur cycle, *150*, 150–151, 188
  from venting clouds, *138*, 140
dinitrogen pentoxide (N_2O_3), *24*
dinosaurs, *11*
diprotic acids, 70, 123
dissolution of CO_2 in oceans, *144*
dissolution of gases, 113, 121–125, *122*, *125*, 131, *144*
Dobson, G. M. B., 184
Dobson units, 165
D-region, *27*, 31
droplets, 93, *93*, 196, 215. *See also* cloud droplets *and* fog
dry adiabatic lapse rate, 47, 62
dry air. *See* air
dry deposition, 80, 100–102, *101*, *148*, 149–151, *150*
dust, 50, 68, *89*, 92–94, 99, 104, 118, 196, 215
dynamic viscosity of air, 97, 100, 196

Earth. *See also* Earth's atmosphere
  aerosols from, 91–94, 104. *See also* aerosols
  age, 69
  albedo, 54
  biological activity, 5, 10, 91
  biosphere, 1–*2*, 7, *7*, 10, 63–64, 72, 96, 143–145, 149
  as a blackbody, *36*, 54
  chemical reservoirs, *2*
  climate, 10–11, 62, 96
  composition of atmosphere, 1–*2*, 21–32, *24*, *27*
  core, 1–2, *2*
  crust, 1, *2*, 7–10, 12, 69, 95, 97–99
  emission temperature, 54–56, 186
  evolution of atmosphere, 1–12, *11*, 21–32, 143, 189–190
  gases from, *11*, *18*, 23–29, *24*, *27–28*, 63–81, *73*, *77–78*, 142–152, *148*, *150*, *166*

Earth (*cont.*)
  hydrosphere, 1, *2*, 143
  life on, 3–8, 11, *11*, 74, 183, 189
  magnetic field, 31
  mantle, 1, *2–3*, 69
  radiation balance, 33–37, *34*, *36*, 54–57, *58*, 121
  solar energy at top of atmosphere, *34*, 35–36
  temperature, 2, 5, 10–11, *27*, 29, 33–35, 54–57, 121, 163, 165, 182, 186
  thermal equilibrium, 33–34, *34*, 51
Earth's atmosphere. *See also* under separate headings
  absorption of radiation, 33–34, *34*, 36–55, *41–42*, *46*, *49*, 57
  aerosols. *See* aerosols
  biological processes, 1, 10, 63–68, 71
  carbon, 1–12, *7*, *9*, *11*, 38, 51, 92, 143–147, *144*
  composition, 1–12, *2*, 21–32, *24*, *27*
  density, 26–29, *28*, 44, 191, 197, 208
  elements, 1–*2*
  evolution, 1–12, *11*, 21–32, 143, 189–190
  greenhouse effect, 10, 54–57, *56*, *58*
  heating, 33–36, *34*, 43–57, *49*, *58*, 165, 186
  heterosphere, *27*, 28–29
  homosphere, 23–26, *27*
  irradiances, *34*, 34–35, *36*, *38*, 39–41, *42*, 43–48, *46*, *52*, 52–56, *56*, 58
  isothermal, 43–44, *46*, 53, 191
  mean free molecular path, *28*
  most abundant elements, 1–*2*
  non-equilibrium, 1, 10
  pressure, 12, 21, *28*, 47, 240
  radiative forcing, 54, 56–57, *58*, 62
  standard, *28*
  temperatures, 5, 10, *27*, 29, 33, 36, 43–44, 47, 54–56, 165, 186
  thermal equilibrium, 33–34, *34*, 51
Earth's crust, 1–*2*, 7–10, 12, 69, 95, 97–99
El Chichon, 180–181, *181*
electrical
  capacitance, 238
  charge, 238
  current, 238
  potential difference, 238
  resistance, 238
electromagnetic (em) radiation, 33–62
electron volt, 238
electrons
  free, *27*, 30–31
  in ionosphere, *27*, 30–31
  unpaired reactive, 32, 74
elemental (*or* black *or* graphitic) carbon, 38, 51, 92, 162
elements, 95, 235–237
emission factors, 156–157, *157*
emission temperature, 54–55, 186
emissivity, 37, 53
energy, 238
enrichment factor for aerosols, 95
entrainment into clouds, *139*, *141*, 141–142
equivalent blackbody temperature, 35
E-region, *27*, 31
escape region (*or* exosphere), *27*, 29, 68
escape velocity, 29–30, *30*
ethylene (C_2H_4), 158
evaporating drops, 96, 102, *139*, 140–141, *141*, 187
evaporation, 33–34, *34*, 51, 101, 155
evapotranspiration, 33–34, *34*
Evelyn, J., 163
exa-, 239
excited electronic states, 29
exosphere (*or* escape region), *27*, 29, 68
extinction coefficient, 37
extinction mass efficiency, 38
eye irritants, 158–159

farad, 238
Faraday constant, 240
femto-, 239
fermentation, 65, 145
fertilizers, *148*, 148–149, 155
film droplets, 93, *93*

first-order photolysis rate coefficient, 58
fixation, 9, 79, *148*, 148–149
fixed nitrogen, 149
flux divergence, 47, 165
fly ash, 87, 92
foam, 93
foaming agents, 171
fog, 121, 129, 156, 199
foraminifera, 8
force, 238
forest fires, 92, 118, *157*. *See also* biomass burning
formaldehyde (HCHO), 4, *18*, *24*, *59*, 65, 73, 76, *144*, 145–146, 159
formic acid (HCOOH), *122*
forward scattering, 50–51
fossil fuel combustion, 7–8, 94, *144*, 146–157, *148*, *150*, *157*, 162, 190
fossil fuels, 7–8, *7*, *58*, 64, 94, *144*, 147, *148*, 148–149, *150*, 156, 190
free electrons, *27*, 30–31
free radicals, 32, 72, 78, 182
F-region, *28*, 31
frequency, 238
fuel NO, 155
fuel rich mixture, 154
fuels, 7–8, *7*, *58*, 64, *144*, 147–156, *148*, *150*, 162, 187, 190
fungi, 91

Gaia hypothesis, 10
gas constant, 44, 121, 129, 191, 200, 240
gas equation. *See* ideal gas equation
gaseous absorption lines, 40
gases. *See also* under separate species
  absorption bands, 40–43, *41*, 62
  absorption lines, 40–41, *41*
  absorption of longwave radiation, 51–54
  absorption of shortwave (solar) radiation, 33, 37–38, 40–43, 45–50, *49*
  absorption spectra, 40–41, *41*
  condensation, 87, 95–97, 98, 138–142, *141*
  detrainment from clouds, *139*
  dissolution in cloud droplets, 121–125, *122*, *125*
  in Earth's atmosphere, 21–32, *24*, *27*
  emission of radiation, 34, *34*, 51–54
  entrainment into clouds, *139*, *141*
  gas-to-particle (g-to-p) conversion, 80, 87, 94, 95–100, *96*, 121
  greenhouse, 5, 11, 56–57, *58*, 147, 163, 164
  reduction of solar radiation by, 40
  residence time (*or* lifetime), 13, 15–19, *18*, 26, 56, 65, 67, 79, 143–147, 151, 187, 209–210
  scattering by, 33–34, *34*, 37–42, 47, *49*, 49–51, *51*, 62, 160–162, 180
  solubility in water, 121–125, *122*, *125*
  in stratosphere, 43, 45, 57, 68, 74, 99–100, *144*, 144–145, 147–148, *148*, 149–151, *150*, 163–184, *182*, 188
  strong electrolytes, 124
  units for, 21–22
gas-to-particle (g-to-p) conversion. *See* gases
gas phase reactions
  and air pollution, 153–160, 162–163
  in evolution of Earth's atmosphere, 1–12
  homogeneous, *73*, *77–78*, 72–79, 81, 111, 137–140, 180
  as sources of aerosols, 94–97, *96*, 98–99
  as sources of chemicals, 63–81, *73*, *77–78*
Geiger counter, 14–15
general circulation models (GCM), 62
geochemical cycles, 69, 143–151
giant aerosols, *86*, 87–88, *93*, 93–94, 97, *103*, *110*
giga-, 239
global
  carbon budget, 143–147, *144*
  nitrogen budget, 143, 149–151, *148*
  pollution, 162–163

global (*cont.*)
-scale, *19*, 92, 96, 153, 162–163, 178, 185
sulfur budget, 143, *150*, 150–151
warming, 54–57, *58*, 121, 163, 187
Gobi Desert, 100, 104
graphitic (*or* black *or* elemental) carbon, 38, 51, 92, 162
Greek smelting, 162
Greenfield gap, 133, *133*
greenhouse (warming) effect, 54–57, *58*, 147
greenhouse gases, 5, 11, 56–57, *58*, 147, 163, 184

Hadley cell, 140
hailstones, 135
half-life of chemicals, 13–15, 17
Halley Bay, 172
halocarbons, 56, *58*, 64. *See also* chlorofluorocarbons *and* hydrofluorocarbons
haloes, 51, 56–57
halogen cycles, 170
Hartley bands, 62
Hartley, W. N., 165, 183
haze (*or* unactivated) drops, 116
hazes, 116, 121, 162
hecto-, 239
helium (He), 2, *18*, *24*, 26–*27*, 29, 68, 236
hemoglobin, 154
Henry's law, 121–126, 142, 199–200
Henry's law constant (*or* coefficient), 121–126, *122*, *125*, 198–199
hertz, 238
heterogeneous, heteromolecular nucleation, 180
heterogeneous reactions, 72, 78, 81, 96
heterosphere, *27*, 28–29
homogeneous aqueous-phase reactions, 72, 78
homogeneous, bimolecular nucleation, 137–140, 180
homogeneous gas-phase reactions, 72–79, *73*, *77–78*, 81, 111, 137–140, 180
homogeneous, heteromolecular nucleation, 180
homosphere, 23–26, *27*
Hudson, *181*
Huggins bands, 62
human health, 153, 162, 183
humans, emergence of, *11*
humidity around clouds, 137, *138–139*
hydrocarbons, *19*, 64–65, 99, 147, 153–155, *157*, 157–158, *160*, 186
hydrochlorofluorocarbons (HCFC), 56, 184
hydrogen (H), 1–*2*, 4, 6, *24*, 26–*27*, 29–30, 74–76, 159, 169–170, 184, 188–189, 193, 203, 206, 231, 236
hydrogen (H_2), 3–4, *18*, *24*, 30, 64, 68, 143, 189, 201, 206
hydrogen bromide (HBr), 68
hydrogen carbonate (H_2CO_3), 8, 69–71, 123–124, 126, 192, 208
hydrogen chloride (HCl), *18*, *24*, 65, 68, 174–176
hydrogen cyanide (HCN), 4, *24*
hydrogen fluoride (*or* hydrofluoric acid) (HF), 68
hydrogen-halogen cycles, 170
hydrogen iodide (HI), 124
hydrogen peroxide (H_2O_2), *18–19*, *24*, 76–77, *122*, 126–128, *127*, 199, 220–223
hydrogen sulfide (H_2S), 10, *18*, *24*, 64, 68, 73, 95, *150*, 150–151, 186–188, 206
hydrolysis, 123, 142
hydrometeors, 131–135, 140
hydronium ion (H_3O^+), 69–71, 126, 198, 218, 222
hydroperoxyl radical (HO_2), *19*, 23–24, *24*, 75–77, *77–78*, *96*, 159, 170, 179, 193, 204–205, 231–234
hydrophobic surface, 142
hydrophylic surface, 142
hydrosphere, 1–*2*, 143
hydrostatic equation, 47, 62
hydrothermal vents, 4

hydroxyl radical (OH), *19*, 23–24, *24*, *59*, 72–75, *73*, *77–78*, 81, *139*, 139–140, *144*, 144–145, 151, 158–159, 170, 186–188, *193*, 193–194
hygroscopic salts, 93

ice cores, 162
ice nuclei, 92
ice particles, 51, 78, 131, 175
ice-water particles, 174
ideal gas equation, 22, 161–162
industrial pollutants. *See* anthropogenic, pollutants *and* polluted air
inertial collisions, 132, *132–133*
infrared (*or* longwave, *or* terrestrial *or* thermal) radiation, 5, 33–34, *34*, 36, *42*, 51–55, *56*, 62, 182, 186
infrared "window," 34, *34*
inorganic aerosols, 98–99
internal combustion engines, 153–154, *See also* automobiles
International System of Units (SI), 22, 165, 238, 240
interstitial aerosol, 112, 140
Io, 201
ion exchange reactions, 8
ionosphere, *27*, 30–31
ion-product constant for water, 70, 240
ions, 8, 10, 30–31, 114, 116
iron (Fe), *2*, 99, 236
iron (Fe^{3+}), 128
iron (III), *127*
irradiance, *34*, 34–35, *36*, *38*, 39–41, *42*, 43–48, *46*, *52*, 52–56, *56*, 58
isooctane (C_8H_{18}), 153–154
isoprene (C_5H_8), 64
isothermal atmosphere, 43–44, *46*, 53
isotopes, 14, 43, 237

jet drops, 93, *93*
joule, 238
Junge, C., 110, 179
Junge distribution, 85–88, *86*
Junge (*or* stratospheric sulfate) layer, 151, 164, *179*, 179, 180–183, *182*

Jupiter, 1, 201
Jurassic, *11*

Kelvin, Lord (Thomson, W.), 110, 113, 142
Kelvin's equation, 113–114, *114*, 115, 142
kilo-, 239
Kirchhoff, G., 37, 61
Kirchhoff's law, 37, 53
Köhler curves, *114*, 115–116, 198
Köhler, H., 115, 142
Koschmeider equation, 160
Krakatoa eruption, 68
krypton (Kr), 2, *24*, 236

landfills, *144*
large aerosols, *86*, 86–87, 94, 102–103, *103*, *110*
latent heat, 33–34, *34*
lead (Pb), 95, 155, 162, 236
leaf litter, 63
leaves, 64
Leighton relationship, 158
lifetime. *See* residence time
light, 33–61, *42*, 160–162, 238, 240
lightning, 4, 79, *148*, 148–149
limestone, 8, 69, 186, 192
Lindemann, F.A., 184
local scale, *19*
log-normal distribution, *90*, 90–91
London, 156, 163
London (*or* classical) smog, 156
longwave radiation. *See* infrared radiation
Los Angeles (*or* photochemical) smog, 64, 155–160, *160*
Loschmidt's number, 22, 240

magnesium (Mg), *2*, 236
magnesium chloride ($MgCl_2$), 90
mammals, *11*
manganese ion (Mn^{2+}), 5, *127*, 127–128, *137*
manure, 65
marine air, 83, *84–85*, 87–*88*, 94–95, 98, *103*, 118, 194

marine boundary layer, 138–140, 194, 200
marine clouds, 95, 98, 99, 119–121, *120*, 137–140, *138–139*
marine organisms, 8
marine sediments, 69
Mars, 1, 8, 185
marshes (*or* swamps *or* wetlands), 64, *144*, 145, *148*, 149, *150*
mean free path (of molecules), 26–31, *28*, 97, 100
mega-, 239
mercury (Hg), 68, 95, 236
mesopause, *27*
mesoscale, *19*
mesosphere, *27*
meteorological range, 160–161
methane (CH_4)
  amount in atmosphere, 3, *144*, 146
  annual increase in troposphere, 146
  in carbon cycle, *58*, *144*, 145–146
  concentrations, 1, 18–19, *24*
  hydrates, *144*
  as oxidant, 74, 159
  oxidation, 3, 74, 76, *144*, 145–146, 187, 192–193, *193*
  radiative forcing by, 56, *58*
  residence time, *18–19*, 145
  sinks, *144*, 144–145, 187
  sources, 3, 64, 68, *144*, 144–145, 159, 163, 185, 189–190, 206
  variability, *19*
methane sulphonate ($(CH_3)_2SOO$), 119
methane sulphonic acid (CH_3SOOH), 124
methanogenic bacteria, 189
methyl bromide, *19*, *24*
methyl chloride (CH_3Cl), *18*, *24*, 64–65, 68
methyl chloroform (CH_3CCl_3), *19*
methyl hydroperoxide (CH_3OOH), *24*, 76, 192–193, *193*
methyl iodide, *24*
methyl mercaptan (CH_3SH), 65
methylperoxyl radical (CH_3O_2), *19*, *24*, 75, 159, 192–193, *193*
methyl radical (CH_3), *24*, 74–75, 159, 175, 193, *193*
micro-, 239
microbes, 64, 91, 149
microscale, *19*
Mie, G., 62
Mie regime, 50
Mie theory, 50
milli-, 239
minerals, 92, 97–98, *106*
mixed nuclei, 118
mixing, 16–17
mixing length, 26
mixing ratio, 21, 23, 131
mixing times, *19*
molar concentration (*or* molarity), 23, 98, 130
molar enthalpy (*or* heat) of reaction, 121
mole, 227, 240
molecular diffusion, 26–28
molecular oxygen. *See* oxygen (O_2)
molecular (*or* Rayleigh) scattering, 41, 50, 161
molecules
  absorption of em radiation by, *42*, 42, 47
  electronic energy, 42
  mean free path, 26–31, *27*, 97, 100
  most probable velocity, 29–30, *30*
  rotational energy, 42–43
  scattering of em radiation by, 37, 41, 50, 160–161
  translational energy, 42–43
  velocities, 29–*30*
mole fraction (*or* ratio), 21–23, 114, 199, 201, 228
Molina, M., 184
monomolecular reaction, 209
Moon, 2
"mother of pearl" (*or* nacreous) clouds *or* Type III PSC, 174–175
Mount Pinatubo, 68, 180–183, *181*
Myakejima, *181*

nacreous (*or* "mother-of-pearl" *or* Type III PSC) clouds, 174–175

nano-, 239
napthalenes, 99
natural gas, 7, 65, *144*, 144–145, 153, 155, 185
natural (*or* anthropogenically undisturbed) stratosphere, 165–171
neon (Ne), 2, *24*, 236
neutral solutions, *136*
newton, 238
nickel (Ni), *2*, 155, 236
nitrate fertilization, 149
nitrate ion (NO_3^-), 74, 94, 98, 149
nitrate radical (NO_3 or $NO_3^\bullet$), *19*, 72–74, 77, 220
nitrate respiration, 64
nitrates, 1, 9, 96, *96*, 98–99, *137*, 163, 149
nitric acid (HNO_3), *24*, 73, *73*, 77, *77–78*, 96, *96*, 98, *122*, 124, 136, 149, 174–176, 180, *182*
nitric acid trihydrate, 174, 180
nitric oxide (NO), 3, *18*, *24*, 29, 61, 64, *73*, 74–76, *77–78*, 79, *96*, 99, *122*, 155, 158–160, 169–171, 187, 193, 195
nitrification, *148*, 149
nitrite ion (NO_2^-), 149
nitrogen (N), *2*, 4, 72, 155, 190, 236
nitrogen (N_2), 1–4, 9–10, 17–18, *18*, 23–24, *24*, 29, 43, 64, 79, 95, *122*, 143, 147–151, *148*, 155, 158, 167, 172–176, 185, 194, 200–201, 227
nitrogen cycle, 147–151, *148*
nitrogen dioxide (NO_2), *19*, 21, *24*, 61, 64, *73*, 73–74, 75–77, *77–78*, 126, 128–129, 149, *157*, 158–160, 162, 170, 175, 187, 190–193, 195, 208
nitrogen fixation, 79, 149
nitrogen oxide (NO). *See* nitric oxide
nitrogen oxides (NO_x), *18–19*, 64–65, 76–77, 99, *148*–149, 153, 155–158, *157*, *160*, 162, 187, 192
nitrogen pentoxide (N_2O_5), 77, 96, *96*
nitrogen trioxide (NO_3), *24*, 74, 77, 220
nitrous acid (HNO_2), *24*
nitrous oxide (N_2O), 1, 18–*19*, 21, *24*, 56, *58*, 64–65, *148*, 149, 163, 200
non-methane hydrocarbons, 64, *78*, *144*
non-seasalt sulfate, 194, 200
normal distribution, 90, *90*
nuclear explosions, 185
nucleation
  of droplets, 111–121, *112*, *120*, 137–140
  heterogeneous, heteromolecular, 180
  homogeneous, bimolecular, 137–140, 180
  homogeneous, heteromolecular, 180
  (*or* nucleus) mode, *86*, 86–87
  scavenging, 111–121, *112*, 125, 134–135, *135*, 140
null chemical cycle, 72
Nyamuragria, *181*

oceans, 3–4, 9–10, 63–64, 69, 71, 79–81, 83, 87–88, 91–94, 96, 99, 104, 118–121, 134, 140, 143–146, 149–151, 185–186, 194–195, 199–200, 224
  aerosols over, 83–88, *84–85*, *88*, 91–94, 97–99, *103*, 103–104, *105*, 118–119, 137–140, 194
  carbon in, *2*, 4, *7*, 64, 143–145, *144*, 147, 200
  formation, 3, 10
  as sink for gases, 9, 72, 81, *144*, 144–145, *148*, *150*
  as source of aerosols, 92–94, *93*, *150*
  as source of gases, 63–68, 71, 80, *144*, *148*, *150*
odd chlorine (ClO_x), 77
odd hydrogen (HO_x), *19*, 26, 75–77, *77–78*, *96*, 159, 193, 205, 232–234. *See also* atomic hydrogen
odd nitrogen (NO_y), 77, 174
odd oxygen (O, O*, O_3), 77, *77*, 169, 171, 183–184, 188. *See also* entries under atomic oxygen and ozone
ohm, 238

oil, 7, 65, 97, 145, 153, 156
oil drilling, *144*, 144–145
optical depth (*or* thickness), 39–42, 45, 50, 54, 104–105, *105*, 180–181, *181*, 191, 196
organics
 aerosol, 50–51, 64, 92–99, *96*, 119
 attenuation of em radiation by, 50
 carbohydrates, 5–6
 carbon, 5–8, 14–15, 51, 64, 92, *144*, 144–145, 185
 in carbon cycle, *144*, 145–147
 composition, 51, 99
 decay, 7
 gas-to-particle conversion, 95–97, *96*, 156
 hydrogen sulfide from, 64
 methane from, 64, 185
 in nitrogen cycle, *148*, 149–151
 from oceans, 91, 94, 119
 pollutants, 155–158, 162
 polycyclic aromatic hydrocarbons (PAH), 99
 radiocarbon dating, 14–15
 solubility, *98*
 sources, 91, 94, 99, 119, 155–158
 in sulfur cycle, *150*, 151–152
oxidants, 73–76, *78*, 125–131, 163
oxidation. *See also* combustion
 capacity of atmosphere, 78, 186
 in chemical circles, *144*, *148*, *150*
 complete, 153–154
 of gases around clouds, *139*
 number (or state), 6, 72, 74, 126, 187, 206
 in photosynthesis, 5–7
 of S(IV) to S(VI), 125–131, *127*
oxygen atom (O, O*), 3–4, 6, *24*, *27*, 29–30, 61, 72–75, *73*, 77, *77*, 155, 158, 167–172, 176, 180, 182–184, 188–189, 193, 195, 202–206, 229–233, 236
oxygen molecule (O_2)
 A-band, *42*–43, 49, 62
 absorption bands, *42*–43, 49, 62
 concentration, 3, *11*, 23–24, *24*
 in Earth's atmosphere, 1–12, *2*, *11*, 17, *24*, 28–30, 23, 143, 154, 185
 evolution in Earth's atmosphere, 1–12
 as oxidant for S(IV), 128–129
 photodissociation, 28
 from photosynthesis, 5–7, 63
 residence time, 17–18, *18*, 26, 191
 solubility, 128
oxygen-17, 43
oxygen-18, 43
ozone (O_3)
 absorption of em radiation by, *42*, 42–43, 45, 47–50, 62
 absorption of ultraviolet radiation by, 11, 45, 74, 165
 bond energy, 81
 Chappuis band, 62
 column abundance, 165–167, 172
 concentrations, 10–11, *11*, *24*, 28, 75, 158, 160, *160*, 164–179, *166*, 187, 195, 202–205, 232–234
 in Earth's atmosphere, 5–6, 10–11, *11*, 29, 43, 45, 49, *73*, 74–76, *77–78*, 81, 158, 160, 164–179, 182–183, 186–188, 195, 202–203, 232–234
 effect on stratospheric temperatures, 165
 electronic transitions, 43
 evolution in Earth's atmosphere, 10–11, *11*
 Hartley band, 62
 hole, 79, 171–178, *173–175*, 183–184
 in homogeneous gas-phase reactions, 72–79, *73*, *77*
 Huggins band, 62
 layer, 6, 165–166, *166*, 179, 184
 as oxidant, *73*, 126–129
 as pollutant, 74, 158, *160*, 187
 reactivity, 81
 residence time, *18–19*
 sinks in troposphere, 20, 72–73, *73*, 75, *77*, 81, 193, 195
 solubility, *122*, 140
 as source of OH, 72–75, *73*, *78*, 140
 sources, 6, 10, *59*, 64, 74–75, 77–78,

78, *139*, 195, 204
in stratosphere, 6, 43, 45, 50, 74, 164–179, *166*, 182–184, 188, 203–205
in troposphere, *19*, 19–20, 43, 49, 56, 72–76, *73*, *77*, 79–81, 140, 164, 193, 195
ozonosphere, *27*, 27–28

Pagan, *181*
PAH. *See under* polycyclic aromatic hydrocarbons
parallel beam radiation, 52
particles. *See* aerosols
pascal, 238
Pavlof, *181*
perfect mixing, 16–17
pernitric acid (HO_2NO_2), *24*
peroxyacetyl nitrate ($CH_3COO_2NO_2$ *or* PAN), *24*, 64, 77, 159, 163
peroxyacetyl radical (CH_3COO_2), 159
peta-, 239
pH
cloud water, 124, 126–128, *134*, 135–136
definition, 71
lemon juice, *136*
milk, *136*
pure water, 71, 135–136, *136*
rain, 136, *136–137*, 195, 213
sea water, *136*
snow, 136
soap, *136*
soft drinks, *136*
stomach acid, *136*
vinegar, *136*
phoretic diffusion, 101–102, 133
photochemical reactions, 4, 10, 29, 33, 57–62, *59–60*, 64, 72–74, 78, 99, 165–169, 189, 202
photochemical (*or* Los Angeles) smog, 64, 155–160, *160*
photodissociation, 3, 28, 167, 201
photolysis, 7, 57, 191
photon, 3, 57
photostationary state, 184
photosynthesis, 5–7, 10, 14, *59*, 63, *144*, 144–145, 147
phytoplankton, 64, 119, 151
pico-, 239
Pinatubo. *See* Mt. Pinatubo
Planck's constant, 240
planetary boundary layer, 79
planets, 1
plants, 5, *11*, 14, 63, 91, 147–151, *148*, *150*, 163, 158, 162–163, 183, 187
polar regions, 32, 83, *84–85*, 165, 172–178, *173–175*, *181*, 187
polar stratospheric clouds (PSC), 172–177, 177, 180–181, *181*, 183–184
polar vortex, 172, *175*
pollen, 91
pollutants, 74, 87, 94, 119, 131, *144*, 146, *148*, *150*, 153–163, *157*, *160*, 187
polluted air, 37, 39, 50, 83, 89, *89*, 94, 96, *103*, 131, 134–135, 153–163, *157*, *160*
polycyclic aromatic hydrocarbons (PAH), 99
polyprotic acids, 142
positive ions, 31
potassium (K), 222, 236
potassium (K^+), 99, *137*
potassium chloride (KCl), 90, 93
potassium-40, 68
power, 238
power plants, 100, *144*, *148*, *150*, 153, 155
Precambrian, *11*
precipitation, 4, 17, 33, 80, 82, 100, 102, 111–113, *112*, *120*, 131–140, *132–133*, *139*, *144*, 146, *148*, 148–149, *150*, 153, 162, 164, 201
precipitation chemistry, 110–142
precipitation scavenging, 10, 17, 80, 102, *112*, 112–113, 131–140, *132–133*, *135*, *139*, *144*, *148*, *150*, 164
pressure, 238
propane, 159
proteins, *148*, 148–149
protons, 14, 32, 70, 212

protozoa, 91
PSC. *See* polar stratospheric clouds
pseudo first-order rate coefficient, 177, 209

quantum yields, 58–60, *60*, 191

radiance, *52*, 52–54
radiant flux from Sun, 34, 54
radiation
  atmospheric, 5–6, 11, 33–62, *34*, 72, 74–78, 82, 104, 165, 178, 183
  attenuation, 33–34, *34*, 36–45, *42*, 50–51, 186
  diffuse, 42, 52
  interactions with trace gases and aerosols, 33–62, *34*
  parallel beam, *38*, 52
  solar (*or* shortwave). *See* solar radiation
  terrestrial (*or* longwave *or* infrared *or* thermal). *See* infrared radiation
radiative energy balance of Earth, 33–62, *34*
radiative forcing, 54–57, *58*
radical(s), 23–24, *24*, 32, *59*, 72
radiocarbon dating, 14–15
radio waves, 30–31
radon (Rn), 68, 69, 186, 236
rainbows, 51
rainwater, 113, 134–137, *135–137*, 194–195, 200
Raoult's law, *114*, 142
rate coefficient for chemical reactions, 14, 240
Rayleigh, Lord (Strutt, J. W.), 61
Rayleigh (*or* molecular) scattering, 42, 50, 161
reflectivity (*or* albedo), 54–57, 61, 119–121, 142, 197
refractive index, 41, 61
refrigerants, 155, 171
regional pollution, 162–163
regional scale, *19*
relative humidity, 90, 115–118, 140
remote sensing, 167, 172, *173–174*
renewal time, 13, 15–17
reservoirs for chemicals, *2*, 143
residence (*or* lifetime)
  aerosols, 100, 102–104, *103*
  atmospheric chemicals, 17–19, *18*, 79, 143–147, 151, 171, 190, 193–194, 200, 205, 209–210, 224, 233
  chemicals in ocean, 65–67, 69
  cloud condensation nuclei, 197
  cloud droplets, 142
  definitions, 15–17
  free electrons, 30
  greenhouse gases, 56
  hydroxyl radical, 72–73
  and spatial and temporal variability, 13, *19*, 19–20
  water vapor, 187
resistance (electrical), 238
resistance (of a surface to uptake), 80
respiration, 7, 63–64, *144*, 147
rhodochrosite ($MnCO_3$), 5
rice fields, 64, *144*, 144–145
Roman smelting, 162
rotational molecular energy, 43
Rowland, S., 184
rubidium (Rb), 61, 237
Ruiz, *181*

Sahara Desert, 99, 104–105, *105*
St. Helens, (Mt.), *181*
saltation, 92
salts, 93, 97–98, *98*, 114–119, *114*, 149
satellite measurements, 104, *139*, 167, 172, *173–174*, 178, *181*, 181–182
saturation vapor pressure, 113–116
scale height, 26, 32, 44, 67, 186, 191, 194, 208, 211, 213
scattering
  coefficient, 37, 160
  efficiency, 50–*51*
  mass efficiency, 38
  molecular, 33–62, *42*, *51*, 160
scavenging. *See* precipitation scavenging *and* clouds, scavenging by
Schwarzschild, K., 53, 62

Schwarzschild's equation, 53
seasalt, 87–88, *88*, 93, *93*, 119, *150*, 150–151
sea-surface temperature, 121
seawater, 64, 66, 92, 94–95, 99, 196
secondary atmosphere, 2
sedimentary rocks, 5, *11*
sedimentation, 97, 100–102, *101*, 196, 215–216
sediments, 69
seeds, 91
shales, 7, *7*
shortwave radiation. *See* solar radiation
Sierra Negra, *181*
silicon (Si), *2*, 99, 237
silver (Ag), 99, 155, 163, 237
SI system. *See* International System of Units
sky, blueness of, 42
smelters, 151–153, 155, 162
smog, 64, 99, 155–160, *160*, 162, 163
smoke, 50, 57, 92, 155, 156, 163
snow crystals, 135
sodium (Na), *2*, 95, 237
sodium chloride (NaCl), 90, 93, 95, 98, 117, 198
sodium ion (Na^+), 99, *137*
soil dust, 50, 92, 97, 104, *106*, 118
soils, 64, 80, 92, 99, *144*, 144–145, 147–148, *148*, *150*, 150–151
solar constant, 35, 191
solar flares, 31
solar irradiance, *34*, 34–35, *36*, 39, 41, 43, 54, 186
solar (*or* shortwave) radiation
  absorption by atmospheric gases, 33–34, *34*, 40–45, *42*, 74
  absorption by Earth, *34*
  attenuation by aerosols, 50–51
  heating of atmosphere by, 45–50, *49*, *58*, 186
  irradiance, 34–47, *42*, 52–58, 186
  and photosynthesis, 6–7, 10, 63
  Rayleigh (*or* molecular) scattering, 41, 50, 161
  reflection, 33–34, *34*, 49, 57, 119, 140, 182, 196
  scattering by aerosols, *34*, 38, 50, 180, 186
  scattering by clouds, 34, *34*, 41, 121
  scattering by gases, 33–62, *34*, *49*, 180
  spectrum, 41–43, *42*, 48, 51
  stratospheric effects, 165–169, 171, 176, 178, 180, 183
solubility of gases, 121–125, *122*, 200
solution chemistry, 111–142
solution droplets, 114–125, *114*, 140, 142
solutions
  chemical reactions, 112, *112*, 119–120, *120*, *125*, 125–131, *127*, 142
  dissolution of gases, 121–125, *122*, *125*
  ideal, 142, 198
  Raoult's law, 142
  units for solute in, 23
solvents, 65, 97, 155, 171
soot, *58*, 92, *See also* black carbon, elemental carbon *and* graphitic carbon
spatial scales of atmospheric species, *19*
specific humidity, 48
spectrum analysis, 61
spores, 91
standard atmosphere, *28*
Stefan-Boltzmann constant, 35
Stefan-Boltzmann law, 35
Stefan, J., 61
Stokes-Cunningham correction factor, 97
Stokes equation, 100
Stokes, G., 110
stratiform clouds, 119–121, *120*, 140
stratopause, *27*
stratosphere
  aerosol "background" state, 180
  aerosols, 164, 179–183, *179*, *181–182*
  anthropogenic perturbations, 80, 164, 171–179

stratosphere (*cont.*)
  chemistry, *59*, 110, 149, 164–184, 188, 202–204
  heating rate, *49*, 49–50
  location, *27*
  lower, 165, 170, 178–179, 183–184, 188
  middle, 170, 203
  ozone, 45, 50, 57–*58*, 62, 74–75, 79, 164–179, *173–175*, 183–184, 188, 202
  ozone "hole", 79, 171–179, *173–175*, 183–184
  sinks for tropospheric chemicals, *144*, 144–145, 147–*148*, 149–150, *150*
  stability, 164
  sulfate (*or* Junge) layer, 151, 164, 179–183, *179*, *182*
  temperatures, *27*, 62, 172, 177–178, 182, 184
  unperturbed (*or* natural), 164–171
  upper, 170, 188, 203
  volcanic eruptions into, 68, 93, 100, 180–183, *181–182*
stromalolites, *11*
Strutt, J. W. *See* Rayleigh, Lord
subduction, 69
sulfate (*or* Junge) layer, 151, 164, 179–183, *179*, *182*
sulfates ($SO_4^=$)
  aerosol, 94, 96–100, 119, 129–140, 149, 151, 162, 179–183, *179*, *182*, 194, *150*, 200
  amount in atmosphere, *150*
  in arctic hazes, 162
  in clean air, *134*, 134–135
  in climate modification, 96
  as cloud condensation nuclei, 96, 119–120, *120*
  concentrations, *24*, 99, *134*, 136–138, *137*, 179, *179*
  formation, 10, 95–96, 100, 119, 134–140, *150*, 150–151, 179–183, *182*
  in hydrometeors, 134–137, *135*, *137*
  in marine air, 64–65, 96, 99, 119, 134, 138–140, *139*, *150*, 150–151, 194–195, 200–201, 211
  in polluted air, 94, 96, 100, 134–135, *134–135*, 162
  in precipitation, 10, 134–137, *134–137*, *150*, 151–153, 194–195, 200–201
  production in cloud water, 125–131, *134*
  radiation effects, 50
  radiative forcing, *58*
  residence time (*or* lifetime), 151, 194, 200, 211
  sinks, *150*, 151–152
  sources, 94, 119, 125–134, *150*, 151–152, 187, 195
  in stratosphere, 151, 164, 179–183, *179*, *182*
  in sulfur cycle, *144*, *148*, 151–152
sulfite ion (SO_3^{2-}), 125–128, 130, 222
sulfur (IV), 125–131, *127*, 199
sulfur (S), 10, 126, 201, 211–213, 237
sulfur (VI), 125–126, *127*, 128–129, 199
sulfur cycle, *150*, 151–152
sulfur dioxide (SO_2)
  amount in atmosphere, *150*
  in arctic haze, 162
  in clouds, 96, *96*, 99, *122*, 156, 125–131
  concentrations, 23–26, *24*, *134*, *150*, 150–151, 156
  conversion to sulfate, 125–131
  deposition velocity, 80, 193
  dissolution (*or* solubility), *122*, 124–131, 156
  emission factors, *157*
  gas-to-particle conversion, 96, *96*, 99–100, *120*, 125–131
  Henry's law constant (*or* coefficient), *122*, 126, 198–199, 220–223
  heterogeneous reactions in cloud water, 96, 99, *120*, 125–131
  hydrated, 126
  residence time (*or* lifetime), 17, *18–19*, 193

sinks, 73, *73*, 81, 138–140, 149–151, *150*
sources, 10, 64, 68, 73, *73*, 94, 100, 119, *150*, 151–153, 155–156, *157*, 180
in stratosphere, 68, 179–183, *182*
in sulfur cycle, *150*, 151–152
venting from clouds, *139*, 139–140
sulfur monoxide (SO), *24*, 182
sulfur oxides (SO_x), 155, 194–195, 211, 213
sulfur trioxide (SO_3), 10, *24*, 179, 195
sulfuric acid (H_2SO_4), 10, *24*, 73, *73*, 76, 87, 98, 126, 136–140, 156–158, *179*, 179–183, *182*
sulfurous acid (H_2SO_3), *24*, 126
Sun
active, *27*
blackbody spectrum, 36, *36*
electrons, 32
faint young, 5, 10
flares, 31
ionizing radiation, 30–31
irradiance, 35–37, 39, 41–43, 52, 56, 62
protons, 32
quiet, *27*
radiation. *See* solar radiation
temperature, 5, 10, 35
ultraviolet radiation, 6, 11, 43, 72, 165–167, 178, 183
X-rays, 31
supersaturation in clouds, 111–118, *114*
supersonic aircraft, 171
surface aerosol distribution plots, 89–90, *89*, 195
swamps (*or* marshes *or* wetlands), 64, *144*, 145, *148*, 149, *150*
synoptic scale, *19*

temperature, 238, *See also* Earth's atmosphere
temperature inversion, 153
temporal scales of variability of atmospheric species, *19*, 19–20
tera-, 239
terminal settling (*or* fall) velocity (*or* speed), 100–101, *101*, 102, 215
termites, 64, *144*
terpenes, 64
terrestrial radiation. *See* infrared radiation
thermal NO, 155
thermal radiation. *See* infrared radiation
thermopause, *27*
thermophoresis, 101, 187
thermosphere, *27*
thiohydroxyl radial (HS), *24*
Thomson, W. *See* Kelvin, Lord
thorium-232, 68, 237
thunderstorms, 164
tobacco burning, 65
torr, 240
transformations of tropospheric chemicals, 64, 72–79, 95–97
transmissivity, 40, 44
transmittance, 40
transport and distribution of chemicals, 79–81, 99, *106*, 188
transportation, *144*
tropopause, *27–28*, 49, 62, 164, 182
troposphere
chemical cycles in, 143–152
free, 79
location, *27*
sink for chemicals, 80–81
source for chemicals, 63–72
-stratosphere exchange, 80, *144*, *148*, *150*
transformation of chemicals in, 72–79
transport and distribution of chemicals in, 79–81, 99, *106*
tundra, 64
turbopause, 28
turbulence, 33, 51, 79, 92
Type I PSC, 174, 180
Type II PSC, 174
Type III PSC (*or* "mother of pearl" *or* nacreous clouds), 174–175

Ulawun, *181*
ultraviolet (UV) radiation, 3, 6, 11, *42*–43, 45, 72, 74, 151, 165–167, 171, 178, 183
unactivated (*or* haze) droplets, 116
Una Una, *181*
universal gas constant, 121, 129, 200, 240
unpolluted troposphere, 75–76
uranium (U), 68, 237
urban air, 38, 40, 87–89, *89*, *103*, 153–163
urea, *148*–149
UV radiation. *See* ultraviolet radiation

variability of atmospheric chemical species, *19*, 19–20
vegetation, 63–64, 80, 92, 118, 100–101, *144*, *148*, *150*
velocity of light, 240
Venus, 1, 8, 12, 185
vibrational molecular energy, 43
viruses, 91
visibility, 82, 96, 100, 116, 153, 160–163
volcanoes
  effects on stratosphere, 100, *181–182*, 182
  emissions from, 2–3, 10, 68–69, 93, *148*, *150*, 164, 180–182, *181*, 201
volt, 238
volume aerosol distribution plots, 89–90, 195

water (H_2O)
  absorption bands, *42*, 42–43, 49
  chemical reactions in liquid, 68–72, 111–142
  condensation, 82, 111–121, 137
  dissociation, 30, 70–71, 123–126
  droplets. *See* droplets, clouds *and* cloud droplets
  on Earth, 1–12, *2*, *18*, 91, 164
  equilibrium vapor pressure above liquid, 113–118, *114*
  formation, 1–2, 11, 153, 201, 227–228
  gas-phase reactions, 73
  ion-product constant, 70, 240
  liquid, 5, 8–10, *18*, 22–23, 82, 111–121, 129–131, 172, 179, 197–198, 215–216, 219–220
  pH, 70–71, 128, 135–136, 195, 212–213, 218
  photolysis, 3, 188
  residence time (*or* lifetime), 17–18, *18*, 104
  soluble salts and organics, *98*
  supersaturation, 95, 111–118, *114*
  vapor, 9, 17, *18–19*, 23–*24*, *42*–43, 48–49, *73*, 101, 104, 111–121, 142, 159, 164, 172, 180, 187
watt, 238
weathering products, 69
wet deposition, 80, 100, 102, *148*, 149, *150*
wetlands (*or* marshes *or* swamps), 64, *144*, 145, *148*, 149, *150*
wettable (hydrophilic) particles, 113, 142
Wien displacement law, 36
Wien, W., 61
windblown spray, 93
wind erosion, 87, 92, 94
windows. *See* atmospheric windows
wood burning. *See* biomass burning
wood molds, 64

xenon (Xe), 2, *24*, 237
X-rays, 31, 61

zinc (Zn), 99, 155, 237